The Standard Encyclopedia of

Carnival Glass

Revised 3rd Edition

Bill Edwards

COLLECTOR BOOKS
A Division of Schroeder Publishing Co., Inc.

The current values in this book should be used only as a guide. They are not intended to set prices, which vary from one section of the country to another. Auction prices as well as dealer prices vary greatly and are affected by condition as well as demand. Neither the Author nor the Publisher assumes responsibility for any losses that might be incurred as a result of consulting this guide.

Printed by IMAGE GRAPHICS, INC., Paducah, Kentucky

Dedication

In Memory of Ellen Edwards
1909–1990

Acknowledgments

Without the help of the following people, the nearly 300 new and replacement photos would not have been possible for me to find in such a short time, and I thank all my "helpers" from the bottom of my heart: Don and Connie Moore, Don and Marian Doyle, Dean and Diane Fry, John and Lucille Britt, Helen Ovellete, Singleton Bailey, Dee Sponser, Joan Nichols, Ray and Verda Asbury, Fred G. Roque, Jr., Carl and Eunice Booker, Lee Markley, Shelli and Clint Arsenault, Eugene R. Grosko, Marie Martin, Lee Briix, Syd and Ola Shoom, Jackie Fisher, the Yohes, Tony Welsh, Pat Davis, Elmer Jenkins, and Mrs. Dow.

And a very special thanks to the San Diego Carnival Club for all their interest and help in finding the glass to be photographed.

Author's Note

I have spent a major portion of my life learning about, living with and writing about Carnival glass. Let me say that aside from the pure pleasure of handling so much of this beautiful glass, I've met so many very wonderful people, I feel my life has been enriched beyond measure by the experience.

I hope my many errors will be excused, and my efforts to aid in the Carnival glass hobby by whatever talents the Lord has given me, remembered. This book is the culmination of all I've learned and a good deal of my soul is here.

Introduction

For the benefit of the novice, Carnival glass is that pressed and iridized glass manufactured between 1905 and 1930. It was made by various companies in the United States, England, France, Germany, Australia, Sweden and Finland.

The iridization, unlike the costly art glass produced by Tiffany and his competitors, was achieved by a spray process on the surface of the glass before firing, thus producing a very beautiful product at a greatly reduced cost, giving the housewife a quality product well within her budget.

In addition, Carnival glass was the last hand-shaped glass mass-produced in America and remains as a beautiful reminder of the glassmaker's skills.

In this volume, we will show the variety of shapes and colors of Carnival glass and will also attempt to define the patterns by manufacturer and for the first time, put the entire field of Carnival glass into one reference book for the collector.

It is my hope that this effort will bring new interest to this truly beautiful glass and stimulate its growth as a collectible; and my only regret is a lack of space to show every known pattern.

The Dugan Story

The Dugan Glass Company of Indiana, Pennsylvania began production on April 14, 1892 calling itself the Indiana Glass Company with Harry White as its president, operating for less than a year before closing. The vacant plant was first leased, then purchased by Harry Northwood in 1895, and for the next two years poured out a stream of Northwood glass until Northwood decided to join the National Glass Combine and moved his main operation to the old Hobbs, Brockunier plant, leasing the Indiana plant to its managers, Thomas E. Dugan and W.G. Minnemeyer who changed the name to the Dugan Glass Company (also called the American Glass Company).

They produced basically the same sorts of glass as Northwood until 1913 when the name was changed again to the Diamond Glass Company with John P. Elkin as president; H. Wallace Thomas, secretary; D.B. Taylor, treasurer; and Ed Rowland, plant manager. During this time, many Northwood molds were reworked, a trademark with a "D" within a diamond registered and most of the Carnival glass from the company was produced under the supervision of Thomas Dugan.

The plant operated until 1931 when it was destroyed by fire and never rebuilt because of the Depression that gripped the country and the industry.

The Fenton Story

First organized in April 1905, the Fenton Art Glass company didn't really materialize until the following July. At that time the glass decorating shop was opened in Martins Ferry, Ohio, in an abandoned factory rented by Frank L. Fenton and his brother, John (who was later to found the famous Millersburg Glass Company).

The next few months were occupied in obtaining financial backers, glass workers, buying land to be plotted into lots as a money-raising venture, and construction of their own plant in Williamstown, West Virginia. At times, everything seemed to go wrong, and it wasn't until 1907 that the company was "on its way."

From the first, the design abilities of Frank Fenton were obvious, and each pattern seemed to bear his own special flair. He (along with Jacob Rosenthal who had come to the Fenton factory after fire had destroyed the renowned Indiana Tumbler and Goblet Company in Greentown, Indiana) was greatly responsible for sensing what the public admired in glass ornamentation.

In 1908, friction arose between the two brothers, and John exited to pursue his dreams in Millersburg, Ohio. By this time, the Fenton process of iridization has taken the mass-scale art glass field by storm and "Carnival glass" was on its way.

For the next 15 years, the Fenton company would produce the largest number of patterns ever in this beautiful product, and huge amounts of iridized glass would be sent to the four corners of the world to brighten homes. While the company made other decorative wears in custard, chocolate glass, mosaic inlaid glass, opalescent glass and stretch glass, nothing surpassed the quality and quantity of their iridized glass. Almost 150 patterns are credited to the company in Carnival glass alone, and many more probably credited to others may be of Fenton origin.

All this is truly a remarkable feat, and certainly Frank L. Fenton's genius must stand along side Harry Northwood's as inspiring.

The Imperial Story

While the Imperial Glass Company of Bellaire, Ohio, was first organized in 1901 by a group of area investors, it wasn't until January 13, 1904, that the first glass was made; and not until nearly five years later the beautiful iridized glass we've come to call Carnival glass was produced.

In the years between these dates, the mass market was sought with a steady production of pressed glass water sets, single tumblers, jelly jars, lamp shades and chimneys, and a full assortment of table items such as salt dips, pickle trays, condiment bottles and oil cruets.

All of this was a prelude, of course, to the art glass field which swept the country, and in 1909 Imperial introduced their iridescent line of blown lead lustre articles as well as the Nuruby, Sapphire and Peacock colors of Carnival glass.

As was quite evident then, as now, this proved to be the hallmark of their production. Huge quantities of the iridized glass were designed, manufactured and sold to the mass marketplace across America and the European Continent for the next decade in strong competition with the other art glass factories. Especially sought was the market in England early in 1911.

In quality, Imperial must be ranked second only to the fine glass produced by the Millersburg company and certainly in design, is on an equal with the great Northwood company. Only the Fenton company produced more recognized patterns and has outlasted them in longevity (the Imperial Glass Company became a subsidiary of the Lenox Company in 1973).

Along the way came the fabulous art glass line called "Imperial Jewels" in 1916. This was an iridescent product often in freehand worked with a "stretch" effect. This is so popular today, many glass collectors have large collections of this alone.

In 1929, Imperial entered the machine glass era and produced its share of what has come to be called Depression glass and in the early 1960's, the company revived their old molds and reproduced many of the old iridized patterns as well as creating a few new ones for the market that was once again invaded by "Carnival glass fever." While many collectors purchased these items, purists in Carnival glass collecting have remained loyal to the original and without question, the early years of Carnival glass production at the Imperial Company will always be their golden years.

The Millersburg Story

Interestingly enough, if John and Frank Fenton hadn't had such adverse personalities, there would have been no Millersburg Glass Company. Both brothers had come to the Martins Ferry, Ohio area in 1903 to begin a glass business in partnership, but Frank was conservative and level-headed while John was brash and eager – a constant dreamer and the pitchman of the family. Each wanted to make a name in the business, and each wanted to do it his way.

By 1908, with the Fenton Art Glass Factory going strong, the clashes multiplied, and John went in search of a location for a plant of his own. He was 38, a huge strapping man with a shock of heavy brown hair and a pair of steely eyes that could almost hypnotize. He would sire five children.

After weeks of travel and inquiry, he came to Holmes County, Ohio, and was immediately impressed with the land and the people. Here were the heartland Americans whose families had come from Germany and Switzerland. They were hard workers like the Amish that lived among them, good businessmen who led a simple life.

Word soon spead that John Fenton wanted to build a glass plant, and the people welcomed him. By selling his interest in the Fenton plant and borrowing, John secured option on a 54.7 acre site on the north edge of Millersburg. Lots were plotted and sold, and ground was broken on September 14, 1908. And like John Fenton's dreams, the plant was to be the grandest ever. The main building was 300' x 100', spanned by steel framing with no center supporting. A second building, 50' x 300' was built as a packing and shipping area as well as a tool shop. The main building housed a 14-pot furnace, a mix room, an office area, a lehr area and a display area for samples of their production.

During construction, gas wells were drilled to supply a source of power and stocks were sold in the company, totaling $125,000.00

On May 20, 1909, the first glass was poured. The initial molds were designed by John Fenton and were **Ohio Star** and **Hobstar and Feather.** While at the Fenton, factory, he had designed others, including the famous "Goddess of Harvest" bowl as a tribute to his wife. The glass that now came from the Millersburg factory was the highest quality of crystal and sample toothpick holders in the **Ohio Star** pattern were given to all visitors that first week.

In addition to the crystal, iridized glass in the Fenton process was put into production the first month in amethyst, green and soft marigold. A third pattern – a design of cherry clusters and leaves – was added and soon sold well enough to warrant new molds, bringing the **Multi-Fruit and Flowers** design into being as a follow-up pattern.

Early in January 1910 the celebrated "Radium" process was born. It featured a softer shade of color and a watery, mirror-like finish on the front side of the glass only; it soon took the glass world by storm and became a brisk seller. Noted glassworker, Oliver Phillips, was the father of the process, and it was soon copied by the Imperial company and others in the area.

In June of that same year, the famous **Courthouse** bowl was produced as a tribute to the town and to the workers who had laid the gas lines to the factory. These bowls were made in lettered, unlettered, radium and satin finish and were given away by the hundreds at the factory, just as the crystal **Ohio Star** punch sets had been given to all the town's churches and social organizations and the famous **People's Vase** was to be made as a tribute to the area's Amish.

During the next years, the Millersburg plant was at its zenith, and dozens of new patterns were added to the line from molds produced by the Hipkins Mold Company. They included the noted **Peacock** patterns as well as the **Berry Wreath, Country Kitchen, Poppy, Diamond, Pipe Humidor** and **Rosalind** patterns. But by late March 1911, Hipkins wanted pay for their work, as did other creditors, and John Fenton found his finances a disaster. The plant was kept producing, but bankruptcy was filed, and Samuel Fair finally bought the works in October, renaming it the Radium Glass Company. In the next few months, only iridized glass in the radium process was produced while John Fenton tried to find a way to begin again.

But time soon proved Fair couldn't bring the factory back to its former glory, and he closed the doors, selling the factory and its contents to Frank Sinclair and the Jefferson Glass Company in 1913. Sinclair shipped many of the crystal and Carnival molds to the Jefferson plant in Canada for a run of production there which included **Ohio Star, Hobstar and Feather** and Millersburg **Flute.** The rest of the molds were sold for scrap despite the success of the patterns with the Canadians who bought the graceful crystal designs for several years. Jefferson's production at the Millersburg plant itself was brief. The known pieces include a 6" **Flute** compote just like the famous Millersburg **Wildflower**, marked "Crys-tal" with no interior pattern.

In 1919, the empty plant was again sold to the Forrester Tire and Rubber Company; the great stack was leveled and the furnace gutted. The Millersburg Glass factory was no more. But as long as one single piece of its beautiful glass is collected, its purpose will stand, and the lovers of this beautiful creation will forgive John Fenton his faults in business and praise his creative genius.

The Northwood Story

An entire book could be written about Harry Northwood, using every superlative the mind could summon and still fail to do justice to the man, a genius in his field. Of course, Harry had an advantage in the glass industry since his father, John Northwood, was a renowned English glass maker.

Harry Northwood came to America in 1880 and first worked for Hobbs, Brockunier and Company of Wheeling, West Virginia, an old and established glass-producing firm. For five years, Harry remained in Wheeling, learning his craft and dreaming his dreams.

In 1886, he left Hobbs, Brockunier and was employed by the Labelle Glass Company of Bridgeport, Ohio, where he advanced to the position of manager in 1887. A few months later, a devastating fire destroyed much of the LaBelle factory, and it was sold in 1888.

Harry next went to work for the Buckeye Glass Company of Martin's Ferry, Ohio. Here he remained until 1896 when he formed the Northwood Company at Indiana, Pennsylvania. Much of the genius was now being evidenced, and such products as the famous Northwood custard glass date from this period.

In 1899, Northwood entered the National Glass combine only to become unhappy with its financial problems, and in 1901 he broke away to become an independent manufacturer once again. A year later, he bought the long-idle Hobbs, Brockunier plant, and for the next couple of years, there were two Northwood plants.

Finally in 1904, Northwood leased the Indiana, Pennsylvania plant to its managers, Thomas E. Dugan and W.G. Minnemeyer, who changed the name of the plant to the Dugan Glass Company (in 1913 the plant officially became known as the Diamond Glass Company and existed as such until it burned to the ground in 1931).

In 1908, Harry Northwood, following the success of his student, Frank L. Fenton, in the iridized glass field, marketed his first Northwood iridescent glass, and Northwood Carnival glass was born. For a 10-year period, Carnival glass was the great American "craze" and even at the time of Harry Northwood's death in 1921, small quantities were still being manufactured. It had proved to be Northwood's most popular glass, the jewel in the crown of a genius, much of it marked with the well-known trademark.

Other American Companies

Besides the five major producers of Carnival glass in America, several additional companies produced amounts of iridized glass.

These companies include: Cambridge Glass Company of Cambridge, Ohio; Jenkins Glass Company of Kokomo, Indiana; Westmoreland Glass Company of Grapeville, Pennsylvania; Fostoria Glass Company of Moundsville, West Virginia; Heisey Glass Company of Newark, Ohio; McKee-Jeanette Glass Company of Jeanette, Pennsylvania; and U.S. Glass Company of Pittsburgh, Pennsylvania.

The Cambridge Company was the "leader" of the lesser companies, and the quality of their iridized glass was of a standard equal to that of the Millersburg Glass Company. Actually, there appears to have been a close working relationship between the two concerns, and some evidence exists to lead us to believe some Cambridge patterns were iridized at the Millersburg factory. The Venetian vase is such an item. Known

Cambridge patterns are:

Horn of Plenty	Inverted Thistle
Sweetheart	Double Star (Buzz Saw)
Buzz Saw Cruet	Near Cut Souvenir
Cologne Bottle	Proud Puss
Forks Cracker Jar	Tomahawk
Inverted Feather	Toy Punch Set
Inverted Strawberry	Venetian
Near-Cut Decanter	

Many of the Cambridge patterns are beautiful near-cut designs or patterns intaglio; and while amethyst and blue are colors rarely found, most Cambridge Carnival glass was made in green and marigold.

The Jenkins Glass Company made only a handful of Carnival glass patterns, mostly in marigold color, and nearly all patterns in intaglio with a combination of flower and near-cut design. Their known patterns are:

Cane and Daisy Cut	Stork Vase	Aztec	Rock Crystal
Cut Flowers	Fleur De Lis Vase	Heart Band Souvenir	Sea Gulls Bowl
Diamond and Daisy Cut	Oval Star and Fan	Lutz	Snow Fancy
Stippled Strawberry		Hobnail Panels	

The Westmoreland Company also made quality Carnival glass in limited amounts. Known patterns are:

Checkerboard	Strutting Peacock
Footed Drape	Pillow and Sunburst
Footed Shell	Shell and Jewel
Basketweave and Cable	Wild Rose Wreath #270
Prisms	Corinth Vt.
Orange Peel	Carolina Dogwood
Fruit Salad	Little Beads

The Fostoria Company had two types of iridized glass. The first was their Taffeta Lustre line which included console sets, bowls and candlesticks; the second was their brocaded patterns which consisted of an acid cutback design, iridized and decorated with gold. These patterns include:

Brocaded Acorns	Brocaded Palms
Brocaded Daffodils	Brocaded Rose
Brodaded Summer Garden	

Heisey made very few iridized items and those found have a light, airy luster. Patterns known are:

Covered Frog	Heisey #357
Covered Turtle	Heisey Flute
Heisey Tray Set	Paneled Heisey

The McKee-Jeanette Company had a few Carnival glass patterns as follows:

The U.S. Glass Company was a combine of 17 companies, headquartered in Pittsburgh, Pennsylvania. Their plants were usually designated by letters, and it is nearly impossible to say what item came from which factory. However, some of the patterns we have classified as U.S. Glass are:

Beads and Bars	Louisville Shrine
Daisy in Oval Panels	Rochester Shrine
Field Thistle	Champagne
Golden Harvest	Shrine Toothpick
New Orleans Shrine Champagne	Palm Beach
Feather Swirl	Vintage Wine
Cosmos and Cane	Butterfly Tumbler

In addition to all the glass produced by these minor concerns, specialty glass houses contributed their share of iridized glass in the large Gone-With-The-Wind lamps, shades, chimneys, etc.; as well as minute amounts of iridized glass from Libby, Anchor-Hocking, Tiffin, Devilbiss, Hig-Bee, and Jeanette are known.

It is not possible in our lmited space to show all these patterns from the smaller makers, but we'll try to give a sampling from many of them.

In addition, many patterns in Carnival glass have not been attributed to definite producers at this time, and so you will find a few items shown where we must say "maker unknown." While we wish this didn't have to be, sooner or later, these too will find their proper place in the history of Carnival glass.

Non-American Carnival Glass Makers

Besides the British and Australians, factories in Sweden, Finland, Denmark, Czechoslovakia, Holland, Mexico, France and Argentina are now known to have produced some amount of iridized glass. While it would be impossible for us to catalogue all these companies, I will list a few of the more prolific and their currently known patterns.

Finland
Finnish Carnival glass wasn't made until the early 1930's and primarily at four factories, including one at Riihimaki which seemed to specialize in copying some best-selling American patterns like Lustre Rose, Four Flowers and Tiger Lily.

Czechoslovakia
Several researchers trace such patterns as Zipper Stitch, The Fish Vase and the Hand Vase to this country's production. If so, they also made some red for I have seen one red Hand Vase! Everything else named is in marigold including the Star and Fan Cordial set.

Sweden
Swedish Carnival glass was in the Varmland region at a factory called the Eda Glassworks with such patterns as Rose Garden, Curved Star, Sunk Daisy and Sunflower and Diamonds. Colors are primarily marigold and blue, and I suspect the Sungold Epergne may have come from this factory.

Argentina
The Regolleau Cristalerias Company is in Buenos Aires, and here a handful of little-known patterns were made including the Beetle Ashtray, and an ashtray I call CR since that is just how it is shaped. I've seen it in both marigold and blue.

Mexico
Recent information into the mysterious patterns of Voltive Light, Ranger and Oklahoma all point to a Mexican production by Cristales de Mexico in Nuevo Leon, Mexico. Each of these patterns is marked with the "M" inside a "C."

In addition, small amounts of iridized glass, often decorated with hand painting, was made in France, Germany, Holland and Belgium. Some of these pieces are very intricate and the design top-notch. I have seen a very beautiful custard glass bowl in a bride's basket that has a candy-ribbon edge; it is iridized and has an enamel design. It was pretty enough to be in a museum, and I've envied the owner ever since I saw it.

English Carnival Glass

When iridized glass caught the buyer's fancy in this country, the major companies (especially Imperial and Fenton) began to ship Carnival glass to England, Europe and Australia, and it wasn't long until the glass house there began to enter the iridized field. In England, the chief producer of this glass was the Sowerby Company of Gateshead-on-Tyne. However, other concerns made some iridized glass, including Gueggenheim, Ltd., of London; and Davisons of Gateshead.

The movement of Carnival glass in England began later than in America and lasted about five years after sales had diminished in this country. Many shapes were made, including bowls, vases, compotes and table pieces. However, water sets and punch sets were pretty much overlooked, and only a few examples of these have come to light. Colors in English glass were mainly confined to marigold, blue and amethyst, but an occasional item in green does appear. The pastels apparently were not popular for few examples exist.

Here then is a list of known English patterns and its volume may surprise many collectors:

African Shield   Hobstar Reversed

Apple Panels
Art Deco
Banded Grape and Leaf
Beaded Hearts
Buddha
Cane and Scroll
Cathedral Arches
   (Hobstar and Cathedral)
Chariot
Covered Hen
Covered Swan
Daisy Block
Daisy and Cane
Diamond Ovals
Diving Dolphins
Fans
Feathered Arrow
Fine Cut Rings
Flute Sherbet
Footed Prism Panels
Grape and Cherry
Heavy Prisms
Hobstar Cut Triangles

Illinois Daisy
Intaglio Daisy
Kokomo
Lattice Heart
Lea (and variants)
May Basket
Moonprint
My Lady's Powderbox
Pineapple
Pinwheel
Pinwheel Vase
Sacic Ashtray
Saint Candlestick
Scroll Embossed Vt.
Signet
Spiralex
Split Diamond
Star
Stippled Diamond Swag
Thistle and Thorn
Tiny Berry Tumbler
Triads
Vining Leaf

Australian Carnival Glass

Just as England caught the "fever," so did the populace of Australia and in 1918, the Crystal Glass Works, Ltd., of Sydney began to produce a beautiful line of iridized glass, whose finish ranks with the very best; mostly in bowls and compotes, but with occasional table items and two water sets known.

Australian Carnival glass is confined to purple, marigold, and an unusual amber over aqua finish and patterns known are:

Australian Swan   Banded Diamonds
Australian Grape   Beaded Spears

Blocks and Arches
Butterflies and Bells
Butterflies and Waratah
Butterfly Bower
Butterfly Bush
Crystal Cut
Emu (Ostrich)
Feathered Flowers
Flannel Berry
Flannel Flower
Golden Cupid
Interior Rays

Kangaroo (and variants)
Kingfisher (and variants)
Kiwi
Kookaburra (and variants)
Magpie
Pin-ups
Rose Panels
S-Band
Sun Gold Epergne
Thunderbird (Shrike)
Waterlily and Dragonfly
Wild Fern

ABSENTEE DRAGON

If you take a close look at this very rare 9" plate, you will see it is very different from the Dragon and Berry pattern it resembles. It is the only one known and while I hate the name, I consider it a top rarity in Fenton glass. It has a Bearded Berry reverse pattern, and the color is excellent. It was, no doubt, an experimental piece.

ACANTHUS

Found both in many shaped bowls and plates, the Acanthus was once considered to be a Millersburg pattern, but old Imperial catalogs have proved its origin as Imperial. The colors are usually marigold or smoke, but I show the pattern in green, and I'm sure amethyst is a strong possibility. The mold work is good, and the glass quality and the iridescence very good.

ACORN (FENTON)

It would be hard to imagine a more naturalistic pattern than this. Realistic acorns and oak leaves arranged in three groupings, filling much of the allowed space, form a pleasing design. Acorn is found on bowls and plates in a wide range of colors including marigold, amethyst, green, red, vaseline, ice blue and iridized milk glass.

ACORN (MILLERSBURG)

Rarely found, this beautiful compote is typically Millersburg in several respects. First, the shape of the stem and the clover-leaf type base are found on several compotes of the Ohio company. Add to that the fine detail of the design, the excellent workmanship, and if that isn't enough, the typical Millersburg colors of green or amethyst, and that should convince anyone. A rare vaseline is known, but a marigold Acorn compote would be a real find.

ACORN BURRS

Other than the famous Northwood Grape and their Peacock at Fountain pattern, Acorn Burrs is probably the most representative of the factory's work, and one eagerly sought by Carnival glass collectors. The pattern background is that of finely done oak bark while the leaves are those of the chestnut oak. The mold work is well done and the coloring ranks with the best. Acorn Burrs is found in a wide range of colors and shapes and always brings top dollar.

ADVERTISING ITEMS

These come in dozens of varieties, several shapes and colors, and it would be impossible in our limited space to show them all. Advertising items were a much-welcomed area of business by **all** Carnival glass makers because they were a guaranteed source of income without the usual cost-lost factor of unsold stock. Also, they could often be made with less care since they were to be given away; old molds could often be utilized, avoiding expensive new molds. Many Northwood advertising items were small plates with simple floral designs.

AFRICAN SHIELD

This small vase-shape originally had a wire flower holder that held the stems of freshly cut blossoms in a neat arrangement. It is 2⅞" tall and 3¼" wide at the top, and while I can't be certain who made it, I believe it is English. Marigold is the only color reported so far.

Absentee Dragon

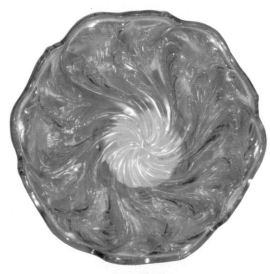

Acanthus

Acorn (Fenton)

Acorn (Millersburg)

Acorn Burrs

Advertising Items

African Shield

11

AGE HERALD

Once thought to be a Millersburg item, we are now convinced the Age Herald bowls and plates were made by Fenton. Found only in amethyst, the Age Herald was a giveaway item from the Birmingham, Alabama newspaper. It has an exterior pattern of wide panels and is a scarce and expensive pattern.

AMARYLLIS

This unusual compote is a treasure for several reasons. First is the size (2¼" tall, 5¼" wide). The shape is roughly triangular, rising from a slightly domed base. The underside carries the Poppy Wreath pattern, and the only colors reported are a deep purple, cobalt blue and marigold.

Amaryllis is a rather scarce Northwood pattern and isn't often mentioned in Carnival glass discussions. It is, however, a unique and interesting addition to any collection.

APPLE AND PEAR INTAGLIO

Like its cousin, the Strawberry Intaglio, this rather rare bowl is a product of the Northwood Company and is seen mostly in crystal or goofus glass. The example shown measures 9¾" in diameter and is 2¾" tall. The glass is ½" thick!

APPLE BLOSSOM TWIGS

This is a very popular pattern, especially in plates where the design is shown to full advantage. The detail is quite nice with fine mold work, much like that of Acorn Burrs. Found mostly in marigold, peach and purple, Apple Blossom Twigs has as its exterior pattern, the Big Basketweave pattern. Shards in this pattern at the Dugan site have been identified.

APPLE BLOSSOMS

Found only in small bowls or plates, Apple Blossoms seems to be an average Carnival glass pattern, produced for a mass market in large quantities. Most often seen in marigold, it is occasionally found in vivid colors as well as pastels, especially white. A quarter-size chunk of this pattern in white was found at the Dugan dump site in 1975.

APPLE PANELS

There is some dispute about the origin of this cute breakfast set; some say Imperial, others England. I personally feel Apple Panels is a British pattern because so many examples come from there. However, the pattern is known in green which is unusual for English glass.

The pattern is all intaglio and no other shapes are known. Green and marigold are the only reported colors.

Age Herald

Amaryllis

Apple and Pear Intaglio

Apple Blossom Twigs

Apple Blossoms

Apple Panels

APPLE TREE

It certainly is a pity Fenton chose to use this realistic pattern on water sets only. It would have made a beautiful table set or punch set. Apple Tree is available in marigold, cobalt blue and white, and I've seen a rare vase whimsey formed from the pitcher with the handle omitted. The coloring is nearly always strong and bright.

APRIL SHOWERS

Like the bubbles in a carbonated soft drink, the tiny beads seem to float over this very interesting vase pattern. Found in all sizes from 5" to 14", April Showers is sometimes found with Peacock Tail pattern on the interior. The colors are marigold, blue, purple, green and white. The top edge is usually quite ruffled in the Fenton manner.

ARCS

This pattern is often confused by beginning collectors with the Scroll Embossed pattern, and it's easy to see why. Perhaps they were designed by the same person since they are both Imperial patterns. Arcs is found on bowls of average size, often with an exterior of File pattern and on compotes with a geometric exterior. The usual colors are marigold or a brilliant amethyst, but green does exist as does smoke.

ART DECO

We don't often see such a plain Carnival pattern, but this one has a good deal of interest, despite its lack of design. The very modern look was all the rage in the Art Deco age so I've named this cute little bowl in that manner. I don't know who the English manufacturer was.

AURORA PEARLS

Here is a piece of beauty no matter if you're a Carnival glass lover or not. It is European, I think, on custard glass, iridized and decorated. If I ever see one for sale, it will be mine for I really am taken by this bowl and only wish it were a compote!

AUSTRALIAN FLOWER SET

Used like the Water Lily and Dragonfly flower set, this Australian beauty has no design except the slender thread border on the bowl's exterior. The iridescence is fantastic, as you can see.

AUSTRALIAN GRAPE

I can't confirm the existence of a pitcher to match this tumbler, but I'd guess one exists. The marigold color is weak, but the mold work is nice. It is from Australia where it is called Vineyard Harvest.

Aurora Pearls

14

Apple Tree

April Showers

Arcs

Art Deco

Australian Flower Set

Australian Grape

AUSTRALIAN SUGARS
These two beauties are kissing cousins and both Australian. The marigold one is called Australian Panels and the purple one Australian Diamonds (not Concave Diamonds as previously listed). Other shapes may exist, but I haven't seen them.

AUSTRALIAN SWAN
Pictured is the Australian Black Swan on a beautiful bowl design. The floral sprays remind us of lily of the valley but are probably some Australian plant. Colors are purple and marigold in both large and small bowls.

AUTUMN ACORNS
Apparently a spin-off pattern from the Fenton Acorn, Autumn Acorns has replaced the realistic oak leaf with the grape leaf used on the Vintage pattern bowls. Found mostly in bowls, an occasional plate is seen in green. The bowls are found in marigold, blue, green, amethyst, vaseline and red.

AZTEC
While McKee didn't make much Carnival glass, the few existing pieces are treasures and certainly the Aztec pattern is one. Known only in a creamer, sugar, rose bowl, tumblers and pitcher, the coloring ranges from a good strong marigold to a clambroth with a fiery pink and blue highlights. Each shape in Aztec is rare and important.

BALLOONS
Balloons is one of those borderline items that could be called either Carnival glass or stretch glass since it has characteristics of both. Found in various shapes, including vases of different sizes and shapes, plates with a center handle, compotes and perfume atomizers. The colors are marigold or smoke. The design is ground through the luster and appears clear; and, of course, there is the stretch effect on many of the pieces.

BAND
Like most of the violet basket inserts, this one relies on a metal handle and holder for any decoration it might have. It was made by Dugan and probably came in marigold as well as the amethyst shown.

BAND OF ROSES
Besides this tumble-up and matching tray, a pitcher and tumbler are known in this pattern. It came from Argentina I understand, but is probably a product of a glass firm in Europe. At any rate, it is pretty and quite rare in any shape.

BANDED DIAMOND
Now known to be an Australian pattern, Banded Diamond is found in water sets and berry sets that are scarce. The colors are a very rich amethyst and a strong marigold.

BANDED DRAPE
Banded Drape is another of Fenton's decorated water sets, but the shape is quite distinctive, being almost urn-shaped. The colors are very beautiful with much luster and are found in marigold, amethyst, cobalt blue, white and ice green. The enameled flower seems to be a lovely calla lily.

Aztec　　　　　　　　　　　　　**Band of Roses**

Australian Sugars

Australian Swan

Autumn Acorns

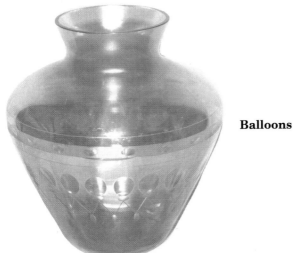

Balloons

Band

Banded Diamond

Banded Drape

BANDED GRAPE AND LEAF

Here is the only water set I've heard about in English glass. As you can see, the design is quite good, and the color is better than average. The only color I've seen is marigold, and at least two sets of this pattern are in American collections.

BANDED RIB

Probably another rather late pattern, this is one of the taller water sets, probably intended for iced tea or lemonade. The only color reported is the marigold, and the only shapes are tumblers and pitchers.

BASKET

Novelty items are a very important area of glass production, and this little item is one of the best known. Standing on four sturdy feet, the Northwood Basket is about 5¾" tall and 5" across. Often the basket is simply round, but sometimes one finds an example that has been pulled into a six-sided shape. Made in a wide range of colors including marigold, purple, vaseline, cobalt, ice green, ice blue, aqua and white. This is a popular pattern.

BASKETWEAVE (NORTHWOOD)

Here's a secondary Northwood pattern found often on bowls and now and then on compotes and bonbons. It's a pretty, all-over filler that does the job.

BASKETWEAVE, BIG

This Dugan pattern is found on the exterior of Fanciful and Round-Up bowls, the base pattern for the Persian Garden two-piece fruit bowls and for vases as shown, as well as a miniature handled basket. Colors are marigold, amethyst, blue, peach opalescent and white.

BASKETWEAVE AND CABLE

Much like the Shell and Jewel breakfast set from the same company, the Westmoreland Basketweave and Cable is a seldom-found pattern and surely must have been made in limited amounts. The mold work is excellent and the luster satisfactory. Colors are marigold (often pale), amethyst, green and rarely white.

BEADED ACANTHUS

It's hard to believe this outstanding pattern was made in this one shape only, but to date I've heard of no other. This milk pitcher, like the Poinsettia, measures 7" tall and has a base diameter of 3¾". It is found mostly on marigold glass or smoke but a very outstanding green exists, and I suspect amethyst is also a possibility. The coloring is usually quite good, and the iridescence is what one might expect from the Imperial Company.

BEADED BAND AND OCTAGON LAMP

Here is a seldom-seen oil lamp that is really very attractive. The coloring is adequate but watery, indicating 1920's production. It was reportedly made in two sizes, 7½" and 9¾", but I can't confirm this. The maker is unknown.

Banded Rib

Basketweave (Northwood)

Banded Grape and Leaf

Basket

Big Basketweave

Basketweave and Cable

Beaded Acanthus

**Beaded Band
and
Octagon Lamp**

19

BEADED BASKET

While the Beaded Basket is quite plentiful, especially in marigold, its origin has always been a little in doubt, and I'm placing it here as a Dugan pattern. Actually, the design qualities are quite good and the mold work superior, so any company could well be proud of these. The colors are marigold, purple, blue and smoke, with the blue most difficult to find. Green may exist, but I haven't heard of one, nor have I had any other pastels reported, except the rare vaseline shown.

BEADED BULL'S EYE

There are several variations of this vase pattern often caused by the "pulling" or "slinging" to obtain height. Nevertheless, the obvious rows of bull's eyes on the upper edge serve to establish identity. Found mostly in marigold, these Imperial vases are not easily found.

BEADED CABLE

The Beaded Cable rose bowl has long been a favorite with collectors for it is a simple yet strong design, made with all the famous Northwood quality. Usually about 4" tall, these rose bowls stand on three sturdy legs. The prominent cable intertwines around the middle and is, of course, edged by beads. Nearly all pieces are marked and are made in a wide variety of colors. Of course, these pieces are sometimes opened out to become a candy dish, like many Northwood footed items.

BEADED PANELS COMPOTE

This little beauty was made by the Westmoreland company and can be found in marigold, amethyst, a rare milk glass with a vaseline finish and the peach opal finish shown. It isn't rare but certainly is a pretty addition to any collection.

BEADED SHELL

Known also in custard glass, Beaded Shell is one of the older Dugan patterns and is found in a variety of shapes, including berry sets, table sets, water sets and mugs. Colors are purple, blue, green, marigold and white with blue and purple somewhat more easily found.

BEADED SPEARS

This previously unlisted Australian water set is a very scarce and beautiful pattern. Besides the peaks of stylized prisms, there are sections of fine file work with unusual circles of plain glass. Other colors and shapes aren't known to me, but certainly may exist.

Beaded Basket

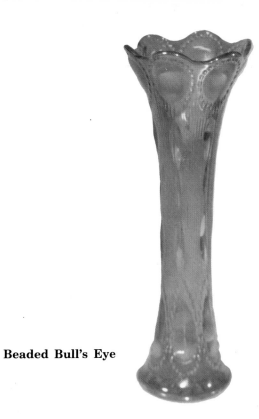

Beaded Bull's Eye

Beaded Cable

Beaded Panels Compote

Beaded Shell

Beaded Spears

BEADED STAR

Most often found on a smallish bowl shape shown, this is a pattern that doesn't receive many raves, nor does it especially deserve them. It is found on rose bowls and a plate shape also in marigold and amethyst.

BEADED SWIRL

This Scandanavian pattern can be found in a covered butter dish, sugar, milk pitcher as well as the compote shown. Colors are marigold and blue, much like the Grand Thistle water set that was made in Finland and this pattern may well be from the same concern.

BEADS

It's really a shame this pattern isn't found more often and is restricted to the exterior of average-size bowls because it is a well-balanced, attractive item, especially on vivid colors. Combining three motifs – daisy-like flowers, petalish blooms and beads – this pattern, while not rare, is certainly not plentiful and is a desirable Northwood item.

BEAUTY BUD VASE (TWIGS)

The most distinctive feature of this vase is, of course, the twig-like feet, intended to be tree roots. These vases are found in sizes from 3½" to 11" tall.

Marigold is the most plentiful color and often only the top shows any hues at all, but the tiny purple version is a beauty and is much sought by vase collectors. Made by Dugan Glass.

BEE ORNAMENT

Like the famous Butterfly Ornaments, the tiny bee is a real rarity and cute as can be (no pun intended). This is the only one I've heard about, but there may be others and even other colors. If so, no one has told me.

BEETLE ASHTRAY

Unusual is the word for this rare ashtray. It was made by the Regolleau Christalerias Company of Buenos Aires, Argentina around 1925. As you can see, the mold work is outstanding, and the cobalt blue coloring is excellent. To date, only two of these have been reported.

BELLAIRE SOUVENIR

This curious bowl measures 7" in diameter and is roughly 2½" deep. The lettering and little bell are all interior work while the fine ribbing is on the outside. Just when this bowl was given, or why it remains is a mystery to me, but I'm quite sure it would have great appeal to the collector of lettered glass. This was made by Imperial.

BELLS AND BEADS

Shards of this pattern turned up in the Helman digs so we know this is another Dugan pattern. Found in small bowls, plates, hat shapes, nappies, compotes and a handled gravy boat, Bells and Beads' colors are marigold, blue amethyst, green and peach opalescent.

BERNHEIMER BOWL

The only difference between this much sought bowl and the famous Millersburg Many Stars pattern is, of course, the advertising center, consisting of a small star and the words: Bernheimer Brothers. This replaces the usual large star and demonstrates how a clever mold designer can capitalize on a good design. While not nearly as plentiful as the Many Stars bowls, the Bernheimer bowls are found in blue only.

Beetle Ashtray

Bee Ornament

Beaded Swirl

**Beauty Bud Vase
(Twigs)**

Beads

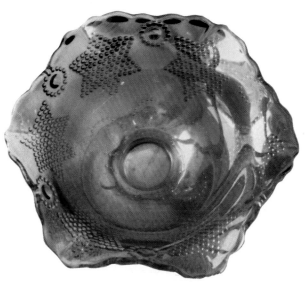

Beaded Star

Bellaire Souvenir

Bells and Beads

Bernheimer Bowl

BIG FISH

If there is a single bowl pattern on which the Millersburg reputation could rest, this wouldn't be a bad choice. Like the Millersburg Peacock, the mold work is outstanding, and the realistic portrayal of the species is exceptional. Each scale, each flower petal is so realistically done that the fish almost seems ready to leap from the glass. Here again, a wide variety of bowl shapes were produced, and it is not unusual to see round, square or three-cornered ones. Often, the amethyst is a little pale, but the green is marvelous and my personal favorite. Again, we find a wide panel exterior with a rayed base.

BIG THISTLE PUNCH BOWL

I could spend pages raving about this very superb Millersburg rarity, but let me simply say I consider it the most beautiful of all the the Carnival glass punch bowls. Two are known, and both are amethyst. One has a flared top while the other is straight up. Needless to say, the glass is clear, the mold work superior and the iridescence beyond belief.

BIRD WITH GRAPES

This unusual wall vase is somewhat similar to the woodpecker vase in concept, and I'd guess it too is a product of the Dugan factory, but I can't be positive. The coloring is a pale marigold with an amber tint.

BIRDS AND CHERRIES

Found quite often on bonbons and compotes, this realistic Fenton pattern is sometimes found on rare berry sets and very rare 10" plates. The birds, five in number, remind me of grackles. I've heard of this pattern in marigold, blue, green, amethyst, white, pastel-marigold and vaseline.

BLACKBERRY (FENTON)

Found often as an interior pattern on the open edge basketweave hat shape Fenton produced in great numbers, Blackberry is a very realistic pattern, gracefully molded around the walls of the hat. It is known in many colors including marigold, cobalt blue, green, amethyst, ice blue, ice green, vaseline and red. The example shown is blue with marigold open edge.

BLACKBERRY (FENTON)

This rare Fenton whimsey is shaped from the two-row open edge basket with Blackberry pattern interior. The vase has been shaped into a 8¼" tall beauty. The only color reported is a beautiful cobalt blue but others may certainly exist.

Big Fish

Big Thistle Punch Bowl

Bird With Grapes

Birds and Cherries

Blackberry

Blackberry Whimsey Vase

BLACKBERRY (NORTHWOOD)

While several companies had a try at a Blackberry pattern, Northwood's is one of the better ones and quite distinctive. The pattern covers most of the allowed space, be it the interior of a 6" compote or an 8½" footed bowl. Combined with the latter is often a pattern called Daisy and Plume. The colors are marigold, purple, green and white.

BLACKBERRY BANDED

Like many Fenton patterns, Blackberry Banded is limited to the hat shape, and without an exterior pattern. These ruffled hats are usually between 3¼" and 3¾" tall, with a base diameter of 2½". Found mostly in marigold or cobalt blue, they are rarely found in green and rare milk glass with marigold iridization.

BLACKBERRY BLOCK

Make no mistake about it – this is a much underrated water set! Just why it hasn't become more treasured by collectors puzzles me, because it is well made, pretty and quite scarce. Found in marigold, green and cobalt blue. White was made too. Manufactured by Fenton.

BLACKBERRY BRAMBLE

A very available pattern on bowls and compotes. Blackberry Bramble is a very close cousin to the Fenton Blackberry pattern but has more leaves, berries and thorny branches. The bowls are rather small with diameters of 6" to 8¼" and the compotes are of average size. Colors are marigold, green and cobalt blue, but others may certainly exist.

BLACKBERRY SPRAY

I've always felt this a poorly designed pattern, but others may disagree. There isn't too much graceful about the four separate branches, and the fruit isn't spectacular, Nevertheless, it can be found on hat shapes, bonbons and compotes, in marigold, cobalt blue, green, amethyst, aqua and red. The aqua does not have opalescence, at least the examples I've seen.

BLACKBERRY WREATH

Apparently one of the early Millersburg patterns, Blackberry Wreath is found more often than not without the radium finish. Oddly enough, the basic pattern is like the Millersburg Strawberry except for the center berry and leaf; however, the mold work is not as distinct, and the glass doesn't have the clarity of the Strawberry pattern. Nevertheless, Blackberry Wreath is a nicely proportioned pattern, especially on the larger bowls. The exterior is usually decorated with a wide panel design and has a many-rayed star on the base, exactly like the Millersburg Peacock and Urn variant.

Blackberry

Blackberry Banded

**Blackberry
Block**

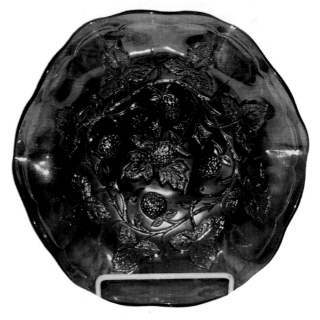

Blackberry Bramble

Blackberry Spray

Blackberry Wreath

BLACK BOTTOM (FENTON)

This very Art Deco little candy jar once had a lid but it has long since been separated. These are shown in the old Fenton ads in a host of colors in stretch glass, but this is the first one I've seen in marigold. The base has a spray-painted finish that has been then fired to keep its black color.

BLOCKS AND ARCHES

Often confused with the very similar Ranger pattern, Blocks and Arches is really an Australian design found only on tumblers and pitchers like the one shown. Colors are marigold and amethyst.

Blocks and Arches

BLOSSOMS AND BAND

Blossoms and Band is not a very distinguished pattern, and its origin is questionable. Found primarily on berry sets in marigold, a car vase is also known, as shown. The design is quite simple with a row of blossoms, stems and leaves above a band of thumbprints and prisms. The color has a good deal of pink in the marigold, much like English glass. The mold work is adequate but far from outstanding. Needless to say, this is not the same pattern as that found on the Millersburg Wild Rose lamps that are often called by the same name. This is possibly an Imperial pattern.

BLOSSOMTIME

Even if this outstanding compote was not marked, we'd surely assign it to the Northwood company because it is so typical of their work. The flowers, the thorny branches (twisted into a geometrical overlapping star) and the curling little branchlets are all nicely done and are stippled, except for the branchlets. The background is plain and contrasts nicely. Blossomtime is combined with an exterior pattern called Wild Flower and is found in marigold, purple, green and pastels. The stem is quite unusual, being twisted with a screw-like pattern. Blossomtime is a scarce pattern and always brings top dollar.

BLUEBERRY

May I say in the beginning, I'm quite prejudiced about this Fenton water set for I think it is outstanding in both design and execution. What a shame it wasn't made in other shapes such as a table set. I've heard of Blueberry in marigold, cobalt blue and white only but that doesn't mean it wasn't made in other colors.

BOOKER MUG

This cutie is in the collection of Carl and Eunice Booker so I've named it after them. It is smaller than most, has a beautiful spray of enameled flowers and an amethyst handle! It is probably European.

BO PEEP MUG AND PLATE

While the Bo Peep mug is simply scarce, the plate is a quite rare item, seldom sold or traded from one collection to another. The color is good marigold and reminds us of that found on most of the Fenton Kitten items. Of course, all children's items in glass were subjected to great loss through breakage, but I doubt if large amounts of the Bo Peep pattern were made to begin with; so of course small quantities have survived.

BORDER PLANTS

Again, here is a pattern long regarded as a Fenton product, but it is actually from the Dugan factory. The colors I've seen are amethyst and peach opalescent, but others may exist. The bowl can be found either flat or footed.

Black Bottom (Fenton)

Blossomtime

Blossoms and Band

Blueberry

Bo Peep Mug and Plate

Booker Mug

Border Plants

BOUQUET

If you look quite closely at this bulbous water set, you'll notice several common devices used by the Fenton company, such as fillers of scales surrounding an embroidery ring. No other shapes exist, and the water set is found in marigold, blue and white. The mold work is quite good, and the colors are typically Fenton.

BOUTONNIERE

This little beauty is a Millersburg product. It usually has a fine radium finish and is most often seen in amethyst. And although I haven't seen one, I've been told there is a variation sometimes found with a different stem and base. Boutonniere is also found in marigold and green.

BOWS AND PANELED DIAMOND

This interesting Fenton pattern is called Paneled and Diamond Bows by Mrs. Presznick. As you can see, it has a roughly geometric panel of diamonds with propeller-like fillers between. These are listed in sizes from 7" to 11". The colors I've heard about are marigold, green, blue, amethyst, and white, but probably there are others.

HEISEY BREAKFAST SET

While it isn't so marked, I'd guess this very attractive marigold breakfast set was a Heisey product. The coloring is very dark and rich and the handles are like those on known Heisey products.

BROCADED ACORNS

The lacy effects of all these brocaded patterns by the Fostoria company are a joy to behold. This was achieved by an acid cutback process, and after the iridescence was fired a gold edging was applied. Found in several shapes, Brocaded Acorns was made in pink, white, ice green, ice blue and vaseline.

BROCADED DAFFODILS

Like the other brocaded patterns, this was made by Fostoria. The shape is a 7½" x 6½" handled bowl in pink. The lovely pattern consists of beautifully realistic daffodils, leaf swirls, and small star fillers.

BROCADED PALMS

Shown is one example of Fostoria's Brocaded series, a large handled cake plate in Brocaded Palms pattern. As you can see, the design is created with an acid cutback effect on the glass before iridization, and the edges are gold trimmed. All the patterns are generally handled in this manner and colors found include ice green, ice blue, pink, white and a lovely and rare rose shade, as well as a rare vaseline color.

Bows and Paneled Diamond

Bouquet

Boutonniere

Heisey Breakfast Set

Brocaded Acorns

Brocaded Daffodils

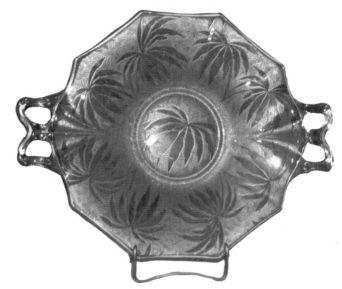

Brocaded Palms

BROKEN ARCHES

Broken Arches is a beautiful geometric Imperial pattern found only in punch sets of rather stately size. Not only is the coloring good, but the mold work is outstanding. The colors are marigold and amethyst. Often there is a silver sheen to the latter which detracts from its beauty, but when a set is found without this gunmetal look the result is breathtaking. The marigold set is more common and sells for much less than amethyst. A green Broken Arches would be a great rarity.

BROOKLYN BOTTLE

I suspect this beautiful 9⅝" cruet may be of European origin. The glass is very thin, and the non-iridized handle and stopper are very attractive amethyst glass.

BROOKLYN BRIDGE

Like the Pony bowl pattern, this beautiful advertising bowl apparently came from the Dugan factory. The only color reported is marigold, and the mold work is outstanding. Brooklyn Bridge is a scarce and desirable pattern. A rare unlettered example is also known.

BULL'S EYE AND LEAVES

Confined to the exterior of bowls and found mostly in green or marigold, this pattern is a trifle too busy to be very effective and is certainly not one of Northwood's better efforts. All in all, there are five motifs, including leaves, beads, circles, fishnet, and a petal grouping. Although each appears on other Northwood products, not in this combination.

BULL'S EYE AND LOOP

This Millersburg pattern is much like its sister design, Tulip Scroll, in size and shape, the example shown being 8½" tall and having a rim diameter of 4". There are four rows of loops with the bull's eye appearing on alternating loops in a staggered manner. The glass is very clear and the radium finish outstanding. Amethyst is the only reported color but others probably exist. Bull's Eye and Loop is a rather rare pattern, and only four examples have been reported to date.

BUTTERFLIES

This outstanding Fenton pattern is found only on bonbons often flattened into a card tray shape. The colors are very good with eight butterflies around the edges and one in the center. The exterior carries a typical wide panel pattern and often is found with advertising on the base. Colors are marigold, cobalt blue, green amethyst and white. At least these are the ones I've heard about.

BUTTERFLIES AND WARATAH

Normally seen in the compote shape, this one has been flattened into a very stylish footed cake stand. The beautiful purple is typical of Australian Carnival glass and can stand with the best. It is also found in marigold.

Bull's Eye and Loop

Broken Arches

Brooklyn Bottle

Brooklyn Bridge

Bull's Eye and Leaves

Butterflies

Butterflies and Waratah

BUTTERFLY

The only shape chosen for this Northwood pattern is the bonbon, and most of the ones I've seen are on amethyst base glass, although marigold and green are found occasionally and pastels have been reported. The pattern shows a lone butterfly in the center of a stippled rays pattern. Not too imaginative but the butterfly shows quite good detail. A variant is known with a ribbed exterior.

BUTTERFLY BOWER

The interior design of a stippled central butterfly flanked by trellis work and flora is hard to see. However, the exterior's S-Band pattern shows quite well. This deep bowl is 6½" across and stands 3" tall. It can be found in marigold and purple. The rim has a bullet edge. Manufactured in Australia.

BUTTERFLY TUMBLER

Shown is one of the most expensive glass tumblers known. This particular example brought more than $4,000.00 in trade and cash a while back, which only proves scarcity plus desire equals big bucks. The Butterfly tumbler is on a pale amberish marigold. The edging design is much like Shell and Jewel, but Butterfly is a U.S. Glass product. A matching footed pitcher has been rumored but not confirmed.

BUTTERFLY AND BERRY

This is certainly one of the Fenton company's prime designs and can be found on a large array of shapes, including footed berry sets, table sets, water sets, a footed hatpin holder, vases, a rare spittoon whimsey and a rare footed bowl whimsey. Colors are marigold, cobalt blue, green, amethyst, white and rarely red.

BUTTERFLY AND CORN VASE

This interesting vase is rare for several reasons, and it is a pleasure to show it here. First is the pattern which has been reported only twice in the past few years. Both examples are identical in size (5⅞" tall and 2¾" base diameter). Secondly, the base color of the glass is vaseline with a marigold finish. While this coloring is found rarely on both Millersburg and Northwood items, I believe the Butterfly and Corn vase to be a product of the latter.

BUTTERFLY AND FERN

For many years, collectors considered this very beautiful water set as Millersburg product and indeed the color and finish rival products of the Ohio company, but Butterfly and Fern is a Fenton item for sure. The mold work is outstanding and the available colors of marigold, amethyst, green and blue are marvelous!

BUTTERFLY AND TULIP

Make no mistake about it, this is a very impessive Dugan pattern. The bowl is large, the glass heavy and the mold work exceptional. Found in either marigold or purple, this footed jewel has the famous Feather Scroll for an exterior pattern. Typically, the shallower the bowl, the more money it brings with the purple bringing many times the price of the underrated marigold.

Butterfly Bower   **S-Band Exterior**

Butterfly

Butterfly Tumbler

Butterfly and Berry

Butterfly and Corn Vase

Butterfly and Fern

Butterfly and Tulip

BUTTERFLY ORNAMENT

I'm told this interesting bit of glass was made as a giveaway item and attached to bonbons, baskets and compotes by a bit of putty when purchasers visited the Fenton factory. This would certainly explain the scarcity of the Butterfly ornament for few are around today. Colors I've heard about are marigold, amethyst, cobalt blue, ice blue, white and green.

BUTTERFLY PINTRAY

While most of these are seen in marigold, often pale and washed out, this example is the pastel version with pink wings and blue body and is a richly iridized beauty. It measure 8" across and is 7" tall. These were made by the same company that made the Tall Hats.

BUTTERMILK GOBLET

Exactly like the Iris goblet also made by Fenton, the plainer Buttermilk Goblet is a real beauty and rather hard to find, expecially in green or amethyst. As you can see, the iridescence is on the interior of the goblet only, and the stem is the same as the Fenton Vintage compotes.

BUTTON AND DAISY HAT

Again, here is an item that has been reproduced in every type of glass known, but the example shown is old and original, and has resided in one of the major Carnival glass collections in the country for many years. Like so many of the miniature novelty items, the coloring is a beautiful clambroth with lots of highlights. Manufacturer unknown.

BUZZ SAW CRUET

This eagerly hunted Cambridge novelty always brings top dollar when it comes up for sale. Found in two sizes, the colors seen are green and marigold. The mold work is fantastic as is the iridescence. Oddly, the base shows a pontil mark indicating the cruet was blown into a mold. Again, these were probably designed as container for some liquid, but just what, we can't say.

CACTUS (MILLERSBURG)

This Rays and Ribbons exterior reminds me of the famous Hobstar and Feather pattern with its incised fans, edged with needles and the hobstars above. In addition, there is a small file filler near the base, merging into a diamond where the feathers meet.

CAMBRIDGE #2351

This Cambridge near-cut pattern was made in a host of shapes in crystal, but in iridized glass only the punch set pieces and a bowl have been reported so far. The punch set was advertised in marigold, amethyst and green, and the one reported bowl is also green.

CANDLE VASE

While I know nothing about this unusually-shaped vase except that it is 9½" tall, I can say with some certainty it was probably shaped in several other ways in many instances. I haven't had any other colors reported, but they may exist.

CANE

One of the older Imperial patterns, Cane is found on wine goblets, bowls of various sizes and pickle dishes. The coloring is nearly always a strong marigold, but the bowl has been seen on smoke, and I'm sure amethyst is a possibility. Cane is not one of the more desirable Imperial patterns and is readily available, especially on bowls; however, a rare color would improve the desirability of this pattern.

Cambridge #2351

Cactus (Millersburg)

Butterfly Pintray

Button and Daisy Hat

Butterfly Ornament

Candle Vase

Buzz Saw Cruet

Buttermilk Goblet

Cane

CANE AND SCROLL (SEA THISTLE)

I'd guess there are other shapes around in this pattern, but the small creamer shown is the only one I've seen so far. As you can see, there are four busy patterns competing with one another but their combination isn't unattractive. The marigold has a reddish hue.

CANNON BALL VARIANT

While the shape of this Fenton water set is the same as the Cherry and Blossom usually found in cobalt, this marigold version has a much different enameled design. The tumblers have an interior wide panel design, and I'm sure this is quite a scarce item.

CAPTIVE ROSE

Captive Rose is a very familiar decorative pattern found in bowls, bonbons, compotes and occasional plates in colors of marigold, cobalt blue, green, amethyst, amber and smoke. The design is a combination of embroidery circles, scales and diamond stitches and is a tribute to the mold maker's art. The roses are like finely stitched quilt work. Manufactured by Fenton.

CARNIVAL BEADS

I have always resisted showing strands of beads before, even though they are very collectible and quite attractive, but when Lee Briix sent this very interesting photo using a Rustic vase as a prop, I changed my mind. Most of these beads were made in smaller glass factories or in Europe.

CAROLINA DOGWOOD

This very interesting Westmoreland pattern is rather hard to find and the few examples I've seen have all been on milk glass base with either a marigold luster or a beautiful bright aqua finish. The design is fairly good, featuring a series of six dogwood sprays around the bowl with a single blossom in the center. Simple but effective.

CARTWHEEL COMPOTE

Not only is this flash-iridized compote marked, it is marked twice – on the base and on the bowl. The color is pale as are most items from the Heisey Company, but quite pretty.

CATHEDRAL (CURVED STAR)

After years of uncertainty and guessing, we've traced this pattern to Sweden and the Eda Glassworks. Several shapes exist including a chalice, pitcher, bowl, flower holder, epergne, compote, butter dish, creamer, open sugar, two-piece fruit bowl, and a rare rose bowl. Colors are marigold and blue.

CENTRAL SHOE STORE

Perhaps this is one of the more scarce Northwood advertising pieces and like most it is a shallow bowl, nearly 6" in diameter. Its lettering reads "Compliments of the Central Shoe Store – Corner of Collinsville and St. Louis Avenues – East St. Louis, Illinois." The floral sprays are much like others used on similar pieces from the Northwood Company.

Central Shoe Store

Carnival Beads

38

Cane and Scroll

Cannon Ball Variant

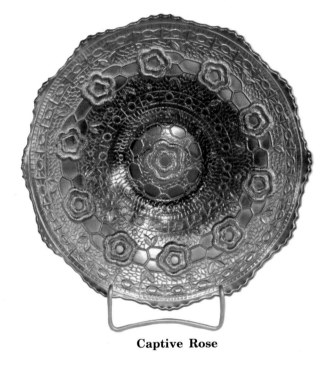

Captive Rose

Carolina Dogwood

Cartwheel Compote

Cathedral (Curved Star)

CHATELAINE

Most of the authorities in the field agree this very rare, beautiful water set pattern is an Imperial item. I'm listing it as a questionable one, however, because I've seen no proof of its origin. Of course, I can't emphasize too strongly its quality or scarcity, and the selling price on the few examples to be sold publicly verify this. The only color I've heard of is a deep rich purple.

CHECKERBOARD

It has been pretty well established that Checkerboard was a Westmoreland product which partially explains its rarity today. I've seen about half a dozen tumblers over the years but the pitcher shown is one of only three known to exist. The color, iridescence and mold work are outstanding. A rare goblet is occasionally found iridized.

CHERRY (DUGAN)

Almost every Carnival glass company produced one or more cherry patterns, and for many years the Dugan Cherry was much confused with the Millersburg Cherry pattern. Of course, there is a considerable difference on comparison and much of the confusion has now been dispelled. The Dugan version is confined to bowls, flat or footed. It has fewer cherries on the branches, no stippling on the branches and less detailed veining on the leaves. The exterior pattern (if there is one) is often Jeweled Heart.

CHERRY (MILLERSBURG)

For many years, Northwood was credited with this beautiful pattern (much as the Fenton company had been for the "Poinsettia" bowl we now know was made by Northwood) but in recent years this error has been rectified. To be sure, the Northwood company did make two Cherry patterns, but the Millersburg Cherry is easily distinguished from either of these since it has more cherries in the clusters, greater variety in the leaf design and appears to be almost "drooping" in appearance.

CHERRY AND CABLE

Sometimes called "Cherry and Thumbprint," this is a very difficult Northwood product to locate, and to date I've seen one tumbler, one pitcher, a table set and a small berry bowl. The pattern is very much a typical Northwood design and reminds one of the famous Northwood Peach, especially in the shape of the butter dish bottom which carries the same exterior base pattern as the Prisms compote. I know of no colors except a good rich marigold, but others may certainly exist.

CHERRY CHAIN

Cherry Chain is a close relative of the Leaf Chain pattern shown elsewhere in this book. Found on bowls, plates and bonbons, this all-over pattern is well done and effective. The colors are marigold, blue, green, amethyst and white. There is an extremely rare example of this pattern on a red slag base glass, and it is the only one I've heard about. There is a variant also.

Chatelaine

Checkerboard Water Set

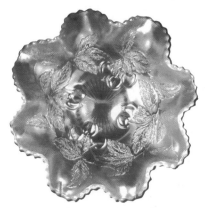

Cherry (Dugan)

Cherry (Millersburg)

Cherry and Cable

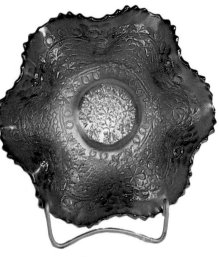

Cherry Chain

CHERRY CIRCLES

This Fenton piece employs a pattern of fruit combined with a scales pattern, the latter being a favorite filler device at the Fenton factory. Cherry Circles is best known in large bonbons, but compotes, bowls and occasional plates do exist in marigold, cobalt blue, green, amethyst, white and red.

CHERRY/HOBNAIL (MILLERSBURG)

On comparison, the interior cherry design varies enough from the regular Millersburg Cherry to show its separate example with the hobnail exterior. These bowls are quite rare and are found in marigold, amethyst and blue and measure roughly 9" in diameter. A 5" bowl can also be found, both with and without the hobnail exterior.

CHRISTMAS COMPOTE

Some dispute has arisen over the origin of this large and beautiful compote, and while some declare it a Millersburg product, I'm inclined to believe it came from the Northwood factory. It is rare and available in both purple and marigold.

CHRYSANTHEMUM

For years I thought this might be an Imperial pattern for it so reminded me of the Windmill pattern, but it really is Fenton. Found on large bowls, either footed or flat, Chrysanthemum is known in marigold, blue, green, ice green, white and a very beautiful red.

CHRYSANTHEMUM DRAPE LAMP

This beauty would grace any glass collection. As you can see, the font is beautifully iridized and has been found in pink glass as well as white. Strangely, most of these lamps have been found in Australia, but the maker is thus far unknown.

CIRCLE SCROLL

Dugan's Circle Scroll is not an easy pattern to find, especially in the water sets, hat shape and vase whimseys. Other shapes known are berry sets, compotes, creamers and spooners. The colors I've seen are marigold and purple, but certainly others may exist with cobalt a strong possibility.

CLASSIC ARTS

What an interesting pattern this is. It is available in a covered powder jar, a rose bowl, a 7" celery vase, and a rare 10" vase. The design is very "Greek" in feeling and the tiny figures quite clear. The green paint gives an antiquing effect which adds greatly to the beauty. It was made in Czechoslovakia.

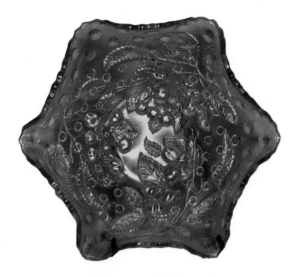

Cherry/Hobnail (Millersburg)

42

Cherry Circle

Christmas Compote

Chrysanthemum

**Chrysanthemum Drape
Lamp**

Circle Scroll

Classic Arts

CLEVELAND MEMORIAL TRAY

Undoubtedly made to celebrate Cleveland's centennial birthday, this cigar ashtray depicts the statue of Garfield, his tomb in Lake View cemetery, the Soldiers and Sailors Monument, the Superior Street viaduct and the Cleveland Chamber of Commerce building. The coloring and iridescence are typically Millersburg, and the mold work compares favorably with that of the Courthouse bowl. A real treasure for any glass collector. Found also in amethyst.

COAL BUCKET (U.S. GLASS)

These rather rare little match holders can be found in both marigold and green and are a miniature collector's dream come true. Few are around, and they bring high prices when sold.

COBBLESTONES

This simple Imperial pattern is not often seen, but certainly is a nice item when encountered. Found on bowls of various sizes – often with Arcs as an exterior pattern. Cobblestones is also found on handled bonbons where a beautiful radium finish is often present, and the exterior is honeycombed! A curious circumstance to say the least! The colors are marigold, green, blue, amethyst and amber. There is also a variant, as well as a rare plate.

COIN DOT (FENTON)

This pattern is fairly common on medium size bowls but can sometimes be found on plates and a handled basket whimsey. The colors are marigold, cobalt blue, green, amethyst and red, but not all shapes are found in all colors. The bowl and rose bowl are the only shapes reported in red.

COIN DOT (WESTMORELAND)

Once these were all thought to be made only by the Fenton company. The variations and some timely research have disclosed that the rose bowl shown, as well as a compote shape and bowl the rose bowl was pulled from, are from the Westmoreland plant. Colors are marigold, amethyst, green, milk-glass iridized, blue milk-glass iridized, teal and aqua opalescent.

COIN SPOT

This undistinguished little Dugan compote holds a dear spot in my heart for it was the first piece of Carnival glass we ever owned. Made in opalescent glass also, in Carnival glass it is found in marigold, green, purple, peach, white and blue. The design is simple, consisting of alternate rows of indented stippled ovals and plain, flat panels. The stem is rather ornate with a finial placed midway down. Often in marigold, the stem remains clear glass.

COLONIAL

Imperial's Colonial pattern is simply one version of wide paneling. The shapes I've heard about are vases, toothpick holders, open sugars, candlesticks and the handled lemonade goblet shown. Colors are marigold, green and purple, usually of the very rich nature.

Coal Bucket (U.S. Glass)

Cleveland Memorial Tray

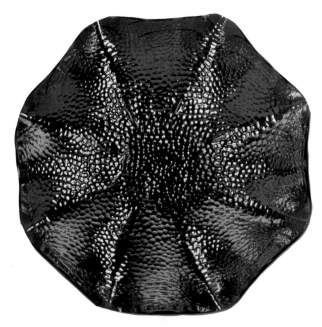

Cobblestone

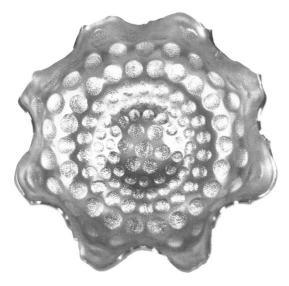

Coin Dot

Coin Dot Rosebowl

Coin Spot

Colonial

COLONIAL LADY

Just look at the color on this rare vase! It certainly rivals Tiffany and only proves Imperial was topped by no one in purple glass. Colonial Lady stands 5¾" tall and has a base diameter of 2¾". I've heard of no other color except marigold, but certainly others may exist. Needless to say, these are not plentiful.

COLONIAL VARIANT

Perhaps this isn't really a variant, but with the lid, it looks different. It was made by Imperial and is found in marigold and sometimes clear Carnival.

COLUMBIA

Made first in crystal, Columbia is found in Carnival glass on compotes and vase shapes, all from the same mold. The coloring is usually marigold, but amethyst and green are known. While the simplicity of Columbia may not be appreciated by some collectors, it certainly had its place in the history of the glass field and should be awarded its just dues. Manufactured by Imperial. Shown in a rare rose bowl whimsey.

CONCAVE DIAMOND

Most of us have seen the Concave Diamond water sets in ice blue and tumblers in vaseline, but I hadn't seen another shape until encountering this quite rare pickle caster in a beautiful marigold. That brought back a memory of finding a small fragment of this pattern in the Dugan shards from the 1975 Indiana, Pennsylvania, diggings, so I would speculate this rare item originated at that factory. Also known are "tumble-ups" in aqua, marigold and aqua opalescent.

CONCAVE FLUTE (WESTMORELAND)

This rather plain pattern is attributed to the Westmoreland company and is found as shown, on rose bowls and pulled out into a vase shape. Colors I've heard about are marigold, amethyst and green but certainly aqua is a possibility.

CONCORD

Even among all the other grape patterns in iridized glass, I'm sure you won't confuse Concord with the others, for the net-like filler that covers the entire surface of the bowls' interior is unique. Found in both bowls and plates, Concord is a very scarce Fenton pattern and is available in marigold, green, amethyst, blue and amber. It is a collector's favorite and doesn't sell cheaply.

CONE AND TIE

The simple beauty of this very rare tumbler (no pitcher is known of now) is very obvious; and while most collectors credit it to Imperial, I cannot verify its maker. The coloring is a very outstanding purple on the few examples known, and the selling price only emphasizes its desirability. Rarely does one of these move from one collection to another.

CONSTELLATION

Constellation is a seldom-seen compote, rather smaller than most. It measures 5" tall and has a bowl diameter that averages 5½" across. The exterior pattern is called Seafoam and I've seen this compote in peach opal, a strange white, over yellow (vaseline) glass and white Carnival. It was made by Dugan.

COOLEEMEE, N.C.

This 9" advertising Heart and Vine plate is a rare one. Only four or five are known and all are marigold. The plate was sold to promote the J.N. Ledford textile mill company of Cooleemee, North Carolina, and undoubtedly was a giveaway item. Today it is a rare and valuable collector's item. Of course, it was made by Fenton.

Concave Flute (Westmoreland)

Cooleemee, N.C.

Colonial Lady

Colonial Variant

Concave Diamond

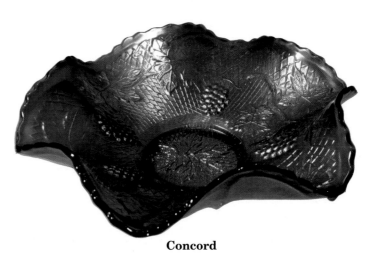

Concord

Cone and Tie

Constellation

Columbia

47

CORAL

I'd guess the same designer who gave us the Peter Rabbit and Little Fishes pattern is also responsible for Coral; the bordering device for all three is almost identical. Coral is found in bowls mostly, but a rare plate and rarer compote are known in various colors, including marigold, blue, green, vaseline and white. Manufactured by Fenton.

CORINTH

Long attributed to Northwood, Corinth is another pattern from the Dugan factory. It is found on bowls, vase and the beautiful banana dish shown, in marigold, green, amethyst and peach opalescent. The example shown is 8¼" long and is iridized on the inside only. A variant is credited to Westmoreland.

CORN BOTTLE

Perhaps the Corn Bottle is not an Imperial product, but the beautiful helios green has always made me feel it was, so while I list it as a questionable Imperial product, I stand convinced it is. Colors found are marigold green, amethyst and smoke (another indication Imperial made it), all of good quality. The iridescence is usually outstanding as is the mold work. It stands 5" tall and usuallly has a cork stopper.

CORN CRUET

Since I first saw this years ago, I've been able to learn little about it. The only color reported is the white Rumor places it in the Dugan line, but I can't be sure.

CORN VASE

This is one of the better-known Northwood patterns, and that's little wonder because it is such a good one Usually about 7" tall, there are certain variants known; especially noteworthy is the rare "pulled husk" varian that was made in very limited amounts. The colors are marigold, green, purple, ice blue, ice green and white The mold work is outstanding, the color superior and the glass fine quality – truly a regal vase.

CORNING INSULATOR

While I've never been very excited by insulators, this one has super color and is marked "Corning – Reg. USA Pat." It is only one of several sizes that can be found.

CORONATION (VICTORIAN CROWN)

This bulbous vase shape stands 5" tall and measures 3" across the lip. It has four Victorian crowns around and suspended coronets between. It is, of course, English, but from what company I can't say, nor can I pin down its use except to call it a vase shape.

COSMOS

Green seems to have been the only color in this Millersburg pattern, and small bowls abound; however, an occasional 7" plate can be found. The radium finish is spectacular, and the mold work is outstanding.

COSMOS AND CANE

What a mystery this pattern has always been to me! Made in white and a soft honey-marigold shade, the shapes are numerous including a berry set, a table set, a water set, a rose bowl, a chop plate, a rare tall compote, a shorter compote and several whimsey shapes. In the whimseys, a rare amethyst color can be found. Shown is a rare oval plate.

Corning Insulator

Corn Cruet

Coral

Corinth

Corn Bottle

Coronation (Victorian Crown)

Corn Vase

Cosmos

Cosmos and Cane

COSMOS VARIANT

Cosmos Variant is a Fenton pattern found in bowls and occasionally plates (a compote has been reported but I haven't seen it). Amethyst is probably the most available color, but Cosmos Variant can be found in marigold, purple, blue, white, iridized milk glass and red.

COUNTRY KITCHEN

Used as both a primary and secondary pattern, this hard-to-come-by design is quite artistic. The berry sets are next to impossible to locate, and the four-piece table sets are quite scarce and expensive. These, of course, are all primary uses; however, Country Kitchen is more often seen as an exterior pattern on the lovely Fleur-de-lis bowls where it compliments the latter perfectly. Millersburg seemed to have a flair for creating outstanding near-cut patterns, and this is certainly one of them. A variant is called Potpourri.

COURTHOUSE

Frankly, I wasn't greatly impressed by the first of these I saw. I'm not quite sure what I'd expected but perhaps it was the smallness of the bowl that disappointed me. Now, however, I wouldn't part with my Courthouse bowl, and I find my admiration for it growing each time I look at it. Known in two shapes – plain or ruffled – and two variations – lettered or unlettered – this 7" shallow bowl typifies the Millersburg technique. The mold work is sharp and detailed, and the glass is very fine and transparent. These things, added to the radium finish, spell quality – a rare thing for a piece of souvenir glass, and points out the John Fenton flair for quality regardless of cost.

COVERED HEN (CHIC)

The Covered Hen is probably the best-recognized piece of English Carnival glass made at the Sowerby Works. Known in a good rich marigold as well as a weak cobalt blue, the Covered Hen is 6⅞" long. The mold work is very good but the color ranges from very good to poor. While reproductions have been reported, my British sources tell me this isn't true, and that no examples have been made since the 1940's.

COVERED FROG

These very realistic covered pieces are 5½" long and stand about 4" tall. They are found in marigold, white, ice blue and the very natural ice green shown. All colors are rare, especially the marigold.

COVERED SWAN

This beautiful English pattern is a companion to the Covered Hen and can be found in both marigold and amethyst. It measures 7¾" long and the neck doesn't touch the back! What a mold maker's achievement! The base is similar to that of the Covered Hen dish. Both are butter dishes.

CRAB CLAW

Crab Claw is indeed a curious near-cut pattern that very seldom enters discussion by collectors, but isn't as readily found as many other geometrics from the Imperial Company. Found on bowls of various sizes, a two-piece fruit bowl and base, and scarce water sets, the pattern is most often found in marigold, but amethyst and green are known in the bowl shapes. The design features hobstars, curving file and diamond devices, and daisy-cut half-flowers, all seemingly interlocking.

Covered Frog

Cosmos Variant

Country Kitchen

Courthouse

Covered Hen

Covered Swan

Crab Claw

CRACKLE

Crackle is very common, very plentiful pattern available in a large variety of shapes, including bowls, covered candy jars, water sets, auto vases, punch sets, plates, spittoons and a rare window ledge planter. Marigold is certainly the most found color, but green and amethyst do exist on some shapes. And often the color and finish are only adequate. Crackle was mass-produced in great amounts, probably as a premium for promotional giveaways, so I suppose we can't expect it to equal other Carnival items.

CRUCIFIX

Now known to be from the Imperial Glass Company, these rare candlesticks are known in marigold Carnival glass and crystal. The Crucifix candlestick is 9½" tall and is very heavy glass.

CRYSTAL CUT

This beautiful compote comes to us from Australia and as you can see, is a very nice geometric design. Other colors and shapes may exist, but if so, I haven't seen them. It measures 7" in diameter.

CURVED STAR (CATHEDRAL)

Here is another view of this Swedish pattern. Shown is the base of the fruit set which, upended, makes a nifty compote as you can see. It is probably used more often in this capacity than its original purpose.

CUT ARCHES

While I'm fairly confident this is another geometric pattern of English origin, I can't remember even seeing another shape in this pattern. As you can see, it is a banana boat of quite heavy glass. The only color reported is the marigold shown, and it has the typical marigold look of English glass. The deeply cut design would certainly lend itself to a water set.

CUT ARCS

While it is most often found on the exterior of 8"–9" bowls, Fenton's Cut Arcs is occasionally seen on compotes of standard size and a vase whimsey, pulled from the bowl shape. On the example shown the edging is a tight candy-ribbon and the interior carries no pattern at all.

CUT COSMOS

How I wish I knew who made this beautiful tumbler! The design is top-notch and certainly deserves full recognition, but the maker remains unknown. The only color reported is the marigold shown, and it has the typical marigold look of English glass. The deeply cut design would certainly lend itself to a water set.

CUT FLOWERS

Here is one of the prettiest of all the Jenkins' patterns. Standing 10½" tall, the intaglio work is deep and sharp and the cut petals show clear glass through the luster, giving a beautiful effect. Most Cut Flowers vases are rather light in color, but the one shown has a rich deep marigold finish. Cut Flowers can also be found in a smoke color.

Cut Arches

Cut Arcs

Crackle

Crucifix

Crystal Cut

Curved Star

Cut Cosmos

Cut Flowers

CUT OVALS

This Fenton pattern falls into the stretch glass field, but because of the etching it is considered Carnival glass. Known in both candlesticks and bowls, Cut Ovals can be found in marigold, smoke, ice blue, ice green, pink, white, tangerine, lavender, red and vaseline. The candlesticks are 8½" tall and the bowls are known in 7", 8", 9" and 10" sizes.

DAHLIA

Make no mistake, Dahlia is an important, often scarce, always expensive, Dugan pattern. The glass is fine quality, the design highly raised and distinct, and the iridescence super. Found only in marigold, purple and white, Dahlia is made only in berry sets, table sets, and water sets, all useful shapes, which probably explains the scarcity due to breakage in use. The water sets are much sought.

DAISY

Found only in the bonbon shape, Fenton's Daisy pattern is very scarce. The pattern is a simple one of four strands of flowers and leaves around the bowl and one blossom with leaves in the center. While marigold has been reported, blue is the color most found, and although I haven't seen amethyst or green, they may exist.

DAISY AND CANE

This rare little decanter is a very unusual design, seldom seen. The coloring is typical of English marigold and has quite good iridescence. I've never seen the stopper so can't verify what it is like. The decanter stands 8" tall and has a 3½" base diameter. I suspect that Sowerby made this, but can't say positively.

DAISY AND DRAPE

This pattern is probably a spin-off of the old U.S. Glass pattern, "Vermont," the main difference being the standing row of daisies around the top edge. Made in most of the Northwood colors, the purple leads the vivid colors, while the aqua opalescent is the most sought of the pastels. White is probably the most available color, but even it brings top dollar.

DAISY AND PLUME

Daisy and Plume is an exterior pattern found on the large Northwood Blackberry compotes as well as the primary pattern of its own compotes and footed rose bowls. It is, of course, a very adaptable pattern and could be used on many shapes. One wonders why it does not appear on table sets or water sets, but unfortunately this is the case. The colors are marigold, white, purple, electric blue, green, peach and aqua.

Dahlia

Cut Ovals

Daisy and Cane

Daisy

Daisy and Drape

Daisy and Plume

DAISY BASKET

Like many other of the Imperial handled baskets in shape and size, the Daisy Basket is a large hat-shaped one with one center handle that is rope patterned. The basket stands 10½" tall to the handle's top and is 6" across the lip. The colors are a good rich marigold and smoke, but be advised this is one of the shapes and patterns Imperial reproduced in the 1960's.

DAISY BLOCK ROWBOAT

Originally used as a pen tray in crystal, this Sowerby product had a matching stand, but I've never seen the stand in Carnival. Daisy Block was made in marigold, amethyst, and aqua in iridized glass. It measures 10½" in length.

DAISY CUT BELL

What a joy this very scarce Fenton pattern is. It was called a "tea bell" in a 1914 catalog, and as you can see, the handle is clear and scored and the Daisy design is all intaglio. The Daisy Cut Bell stands 6" tall and is found in marigold only. It is a four mold design and has a marking inside "PATD APPLD."

DAISY SQUARES

When I first showed this pattern in my Millersburg book 15 years ago, I truly believed it came from that company, but over the years I've grown skeptical. At any rate, it can be found in compotes, rose bowls and goblets, all from the same mold and in marigold, amethyst, a light airy green and a strange amber base glass with over coloring of green.

DAISY WEB

What a hard-to-find pattern this is. Only the hat shape is known, and the only colors I've seen are amethyst, blue and marigold. From the exterior design of Beaded Panels we know this pattern came form the Dugan company, so peach is a definite possibility.

DAISY WREATH

Apparently Westmoreland decided this pattern looked best on a milk glass base for I haven't heard of it any other way. The 9" bowl in marigold on milk glass is quite scarce. Equally scarce is the aqua blue on milk glass shown.

DANCE OF THE VEILS

I've heard of three of these beauties by the Fenton Company in iridized glass, and they are truly a glassmaker's dream. Marigold is the only Carnival color although crystal, pink, green, custard and opalescent ones are made.

Daisy Web

Daisy Basket

Daisy Block Rowboat

Daisy Cut Bell

Daisy Squares

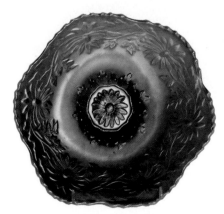

Daisy Wreath

Dance of the Veils

DANDELION

I have serious misgivings about labeling both of the pictured items as the same patterns, but since this is how they are best known, I will yield to tradition. Actually, the mug is certainly a dandelion pattern that is quite rare and popular, especially in aqua opalescent. Occasionally a mug bears advertising on the base and is called the "Knight Templar" mug; it brings top dollar!

The water set, while known as Dandelion, is not the same design, but nonetheless is an important Northwood pattern. It is found in marigold, purple, green, ice blue, white and a very rare ice green. The tankard pitcher is regal.

DEEP GRAPE COMPOTE

Ever notice how few compotes were made by Millersburg? This is one of the rarer ones, found in cobalt, amethyst, green, marigold and vaseline. It stands about 7" high and the stem is paneled. The grape clusters and leaves are richly detailed and the mold work is outstanding. Shown is a rare rose bowl shape.

DIAMOND AND DAISY CUT

This beautiful pattern is found on 10" vases, compotes and rare water sets (which have been mistakenly labeled Mayflower). Found primarily in a good rich marigold, the compote has been in amethyst so perhaps other colors exist.

DIAMOND AND DAISY CUT VARIANT

While not exactly like the water set with the same name, this punch bowl is near enough to be termed a variant, resembling the compote more nearly. It is the only example thus far reported and I know of no base or cups but certainly there must have been.

DIAMOND AND FAN (MILLERSBURG)

This is a beautifully designed exterior pattern, always found with the Nesting Swan bowl in Carnival glass. As you can see, the pattern literally covers all available space but doesn't seem busy.

DIAMOND AND FILE

I'm not at all sure this pattern is Imperial, but list it here as a possible one. It hasn't a great deal of imagination, but may have been intended as giveaways or quick sales. The small bowl shown is quite deep, possibly intended as a jelly holder. The color is adequate, and I've heard of these in smoke as well as marigold.

DIAMOND AND RIB

Diamond and Rib is a Fenton pattern that has caused great confusion on the collecting world. This is because the jardiniere shape was long suspected to be a Millersburg product. This size has been found in large vases just like the smaller ones, so that we know there were two sizes of molds used. Colors are marigold, green, amethyst, smoke and blue.

DIAMOND AND SUNBURST

Known in bowls, decanters, goblets and the rare oil cruet shown, Diamond and Sunburst is an Imperial pattern. Colors are marigold, green, purple and amber.

Diamond And Fan (Millersburg)

Diamond and Daisy Cut Variant

Dandelion

Deep Grape Compote

Diamond and Daisy Cut

Diamond and File

Diamond and Rib

Diamond and Sunburst

DIAMOND FLUTES

This pattern is shown in a 7¼" parfait glass. The coloring is a good marigold at the top, running to clear at the base. Other shapes are not reported but may exist.

DIAMOND FOUNTAIN

This beautiful cruet has now been traced to the Higbee company and appears in their 1910 ads as a pattern called Palm Leaf Fan in crystal. As you can see, the mold work is quite good and the color adequate.

DIAMOND LACE

There seems to be a great deal of doubt about the maker of this beautiful water set and some credit it to Imperial, while others proclaim it to be a Heisey product. I personally lean toward Imperial as the maker because of the beautiful coloring, the mold sharpness and the clarity of glass. In most ways, it greatly resembles the Chatelain water set or the Zippered Heart berry set as to coloring and mold work. Diamond Lace is found in berry sets and water sets in colors of marigold and purple, but white has also been reported.

DIAMOND OVALS

Known in only two shapes thus far, a creamer on a foot and a stemmed compote, Diamond Ovals is a Sowerby product. The creamer is one of the smaller ones, measuring 4¾" tall. The only color reported is marigold, but certainly others may exist.

DIAMOND PINWHEEL

Perhaps there are other shapes in this pattern, but if so, I haven't seen them. As you can see, it is a simple geometric design, but very pleasing. The glass and luster all have good quality and certainly add to any collection.

DIAMOND POINT

While this Northwood pattern is quite typical of most vase patterns, it has a certain distinction of its own. While refraining from "busyness," it manages an interesting overall pattern. It is found in purple, blue, aqua opalescent, green, marigold, peach and white and is normally of standard height, being 10" to 11" tall.

Diamond Flutes

Diamond Fountain

Diamond Lace

Diamond Ovals

Diamond Pinwheel

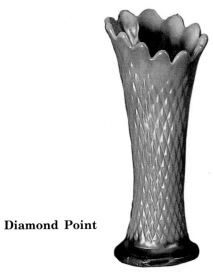

Diamond Point

DIAMOND POINT COLUMNS

I haven't heard of this Imperial pattern in any other color than a good rich marigold, but that certainly doesn't mean they do not exist. The shapes known are compotes, plates, vases, table sets, and powder jars (which are rare). While the design of alternating rows of plain and checkered panels is quite simple, the overall appearance is quite effective. It is also found in small bowls.

DIAMOND POINTS BASKET

Although the dimensions are nearly the same as Northwood's Basket, the Diamond Points Basket is a much rarer pattern. And while some collectors have credited the Fenton company with this much-sought novelty, I lean toward Northwood as the maker. The colors found are marigold, purple and a radiant cobalt blue. The mold work is sharp and precise and the iridescence heavy and outstanding. I have heard of less than a dozen of these baskets, so lucky are you if you own one.

DIAMOND PRISMS

I've seen this same compote shaped with the sides squared, and it was also on the same shade of marigold with much amber in the tint.

DIAMOND RING

This typical geometric design of file fillers, diamonds and scored rings is found on berry sets as well as a fruit bowl of larger dimensions. The usual colors of marigold and smoke are to be found and in addition, the fruit bowl has been reported in a beautiful deep purple. While Diamond Ring is not an outstanding example of the glass maker's art, it certainly deserves its share of glory. Manufactured by Imperial.

DIAMONDS

Most collectors are familiar with this pattern in the water sets but it does appear on rare occasions in a punch bowl and base (no cups have yet come to light). Like its cousin, Banded Diamonds, the design is simple but effective. The glass and iridescence are very good and all have the "Millersburg look."

DIVING DOLPHINS (ENGLISH)

For many years, this scarce item was credited to the Imperial company, but we know now it was a Sowerby product. The interior carries the Scroll Embossed pattern, and Diving Dolphins is found in marigold, blue, green and amethyst with the latter easiest to find. It measures 7" across the bowl and may be round, ruffled, turned in like a rose bowl, or five-sided.

Diamond Point Columns

Diamond Points Basket

Diamond Prisms

Diamond Ring

Diamonds

Diving Dolphins

DOGWOOD SPRAYS

This is a pattern so very familiar to all Carnival collectors, for there are probably more bowls with variou floral sprays than any other design. Dogwood Sprays, while well done, is not extraordinary in any sense of th word. Found on both bowls and compotes, the color most seen is peach but purple does turn up form time to time Manufactured by Dugan.

DOLPHINS COMPOTE

If I had to pick my very favorite Millersburg pattern, this would be it hands down. Not only is it unique i the entire field of Carnival glass by virture of its exterior design, but combined with the Rosalind interior, a almost perfect coordination is brought about. It simply looks good from any angle. Add to this the fine colorin and the radium finish and the Dolphins Compote is a true work of art, one that any company could be prou of.

DOUBLE DOLPHIN

Some people will probably question this pattern appearing here for it is rightly considered stretch glass, bu the mold work of the dolphins tends to bridge the gap between the stretch and Carnival glass so I've shown i here. This Fenton pattern was made in many shapes, including bowls, compotes, vases, covered candy dishes candlesticks and cake plates. Colors are ice blue, ice green, white, pink, red, topaz, tangerine and amethyst.

DOUBLE DUTCH

While it's difficult to say whether this pattern came before or after Windmill and the NuArt plate, it is quit obvious they were all the work of the same artist. All depict various rural scenes with trees, a pond and bridge This Imperial pattern is featured on a 9½" footed bowl, usually in marigold but occasionally found in smoke, gree and amethyst. The coloring is superior and the mold work very fine. The exterior bowl shape and pattern is muc like the Floral and Optic pattern.

DOUBLE FAN TUMBLER

Please notice the very appealing shape of this rare, rare tumbler (two known), for it is very balanced as is th design. And while the coloring is only so-so, the glass is very clear and the mold work excellent. The Double Fa tumbler stands 3¾" tall and is 2½" across the diameter of the base. Certainly a complete water set would be treasure indeed, regardless of the manufacturer.

DOUBLE LOOP

This Northwood pattern is found on creamers, as shown, and an open sugar with a stem. Often these ar trademarked, but not all are. The colors known are marigold, green, purple, aqua opalescent and cobalt with th latter most difficult to find.

Dogwood Sprays

Dolphins Compote

Double Dolphin

Double Dutch

Double Fan Tumbler

Double Loop

DOUBLE SCROLL

The Double Scroll pattern was Imperial's try at an Art Deco pattern in iridized glass and can be found on candlesticks, a dome-based console bowl, and a punch cup in marigold, green, amethyst, smoke and red (not all colors in all shapes). The candlesticks measure 8½" tall and the oval-shaped console bowl that went with them is 10½" x 8½" in diameter and stands 5" tall. The scrolls are of solid glass. The glass is thick and fine and the coloring is excellent.

DOUBLE STAR

In reality, this is the same pattern as the Buzz Saw cruet, but for some reason it has been called by this different name. Found in water sets and a rare spittoon whimsey formed from a tumbler, this Cambridge pattern is seen mostly in green, but marigold and amethyst can be found. I consider Double Star one of the best designed water sets in all of Carnival glass.

DOUBLE STEM ROSE

Like other patterns made at the Dugan plant, this pattern has mistakenly been attributed to the Fenton company. Double Stem Rose is found on dome-footed bowls of average size in marigold, blue, green, lavender, amethyst and white. The design is interesting and well executed.

DRAGON AND BERRY (STRAWBERRY)

Obviously a companion piece to well-known Dragon and Lotus pattern, Dragon and Berry apparently didn't have the popularity of its cousin for few examples are to be found. Known in both footed and flat based bowls and in colors of marigold, green, and blue, this Fenton pattern is a much-sought item.

DRAGON AND LOTUS

Probably one of the best known of all Carnival glass patterns, Dragon and Lotus was a favorite of all Fenton patterns produced over several years. Shapes are flat or footed bowls and rare plates. Colors are marigold, green, blue, amethyst, peach opalescent, iridized milk glass, aqua opalescent, vaseline opalescent, red and the pastels.

DRAGON'S TONGUE

Known mostly in light shades in a milk glass with marigold iridescence, Dragon's Tongue is also found on a large footed bowl that has a diameter measurement of 11". The only color reported on the bowl shape is marigold, but certainly other colors are possible, especially Fenton's famous cobalt blue.

DRAPERY

This is a very graceful Northwood pattern, especially when the vase has been pulled out and down to form a lovely candy dish. One is instantly reminded of great folds of soft satin draperies in a theater, especially in the vase shape, most of which are 10" to 11" in height. The colors are marigold, peach, blue, purple, green, ice blue and ice green. Shown is a rose bowl shape.

Drapery

66

Double Scroll

Double Star

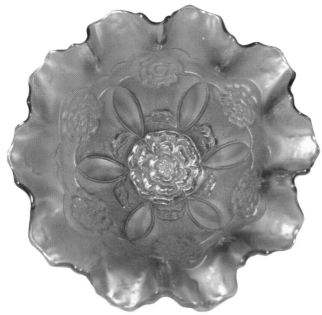

Double Stem Rose

Dragon and Berry

Dragon and Lotus

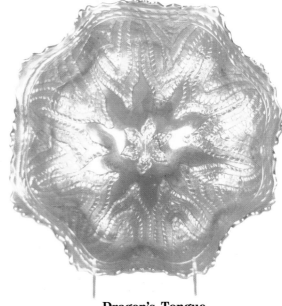

Dragon's Tongue

DUGAN'S VINTAGE VARIANT

I certainly wish we could avoid all the variants, but for the life of me, I don't know what else to call this ver rare plate. It is one of five or six known, is much different in arrangement of grape clusters and the tendrils of the leaf center and the edging is different. It has a 1" collar base and is found in amethyst as well as the marigol shown.

ELEGANCE

What a beauty this very rare plate is. I suspect the maker was either Fostoria or Cambridge but can't be sur at this time. An 8½" bowl is known in amethyst and the pattern reminds me of Hearts & Flowers somewhat. S far only the two pieces have been reported but surely there are others.

ELKS (FENTON)

Found in bowls, plates and the quiet rare bell shapes, the Fenton Elk pieces were produced to be sold at Elk conventions in Detroit (1910), Atlantic City (1911), and Parkersburg (1914). The colors found are a very rich cobalt blue and a sparkling emerald green with heavy luster. Naturally the plates are more sought than the bowl and the bells are highly prized treasures.

ELKS (MILLERSBURG)

The Millersburg Elks bowl or "two-eyed elk" is the best of all the elks bowls or plates. It is somewhat large than the Fenton pieces. The only color I've seen is a rich amethyst with beautiful radium finish. In addition t bowls, paperweights have been found.

ELKS NAPPY (DUGAN)

For many years I thought this beautiful nappy (two known) was a Millersburg product but recent informatio assures us it came from the Dugan factory. As you can see, it is a real beauty with stunning color and fine mol work. The only color is purple.

EMBROIDERED MUMS

Embroidered Mums is a rather busy pattern, saved from being overdone by a balancing of its parts. A clos cousin to the Hearts and Flowers pattern, Embroidered Mums is found on bowls and bonbons on a stem. Man of the pieces of this pattern carry the Thin Rib as a secondary pattern and perhaps a plate will someday be found At least, it gives us something to hope for, because many Northwood patterns with this exterior are found i plates.

EMU

The Emu bowl is one of the better Australian patterns and is available on 4½" and 10½" bowls as well as larg compotes and footed cake stands. The color most found is purple, but marigold and the amber over aqua bas glass are found rarely. Some call this pattern Ostrich, but the Ostrich lives in Africa and the Emu is a Australian bird with a similar appearance.

ENAMELED CARNIVAL GLASS

Toward the latter years of the Carnival glass "craze," patterns become simpler, and in order to give th customer something different, items (especially water sets and bowls) were marketed with hand-painted ename work on them. Most of these were simple floral sprays, but occasionally an interesting fruit pattern emerged, lik the scarce water set shown here. Today, these seem to be gaining in popularity and certainly deserve a place i Carnival glass history. This is a Northwood pattern called Enameled Grape.

Elks Nappy
(Dugan)

Dugan's Vintage Variant

Elegance

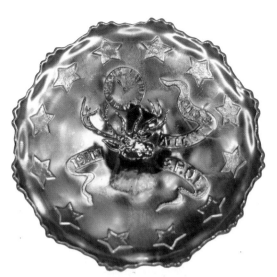

Elks (Fenton)

Elks (Millersburg)

Embroidered Mums

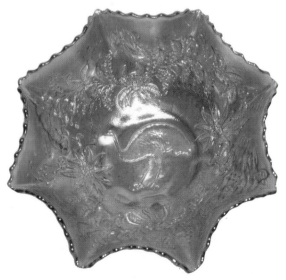

Emu

Enameled Carnival Glass

ENAMELED PRISM BAND

This beautiful Fenton tankard water set is a standout in the series of enameled water sets. It can be found in marigold, blue, green (rarely) and a very impressive and scarce white. Also the floral work may vary slightly from one item to another.

ENGLISH HOB AND BUTTON

This pattern has been reproduced in the last few years in this country, especially on tray and bowl shapes in an odd shade of amberish marigold on very poor glass. The English version shown is another thing, however. The glass is clear and sparkling. The shapes are bowls, mostly, in marigold, green, amethyst and blue.

ENGRAVED GRAPE

This lovely etched pattern is from the Fenton factory and has been seen on many shapes including vases, candy jars, tall and squat pitchers, tumblers, juice glasses and a tumble-up. Colors are marigold and white.

ENGRAVED ZINNIA

While the engraving of tall and short-stemmed flowers is rather difficult to see, the tumbler has good luster and the shape is quite nice. Has anyone seen the pitcher?

ESTATE MUG

This rare little mug appears to belong in the late Carnival era, but nevertheless is a much-in-demand item especially to mug collectors. Most of these mugs are souvenir items, and the one shown is no exception; it bears the inscription: "Souvenir of Gant, New York." The coloring is a pale marigold and the iridescence is only so-so. It measures 3" tall.

EVELYN

This very rare bowl pattern was never put into production and according to Marvin G. Yutzey, former Design Director at Fostoria, all other pieces of the trial run were destroyed. This sole survivor was owned by Evelyn Guest of Canton, Ohio, and now belongs to her daughter, Judy Rhodes. I've named it for Evelyn who was a gracious lady and a true friend. The bowl dates from the late 1930's or early 1940's.

FAN

Despite the fact most collectors have credited this pattern to Northwood. I'm really convinced it was a Dugan product. In custard glass it has been found with the well-known Diamond marking. Of course, many more shapes of the Fan pattern were made in custard. In Carnival glass, the availability is limited to the sauce dish and an occasional piece that is footed and has a handle.

The colors seen are marigold, peach and purple. Peach is the most available color.

FAN-STAR (MILLERSBURG)

Except for the base, this very unusual Millersburg exterior is completely plain. The base has a very odd star bursting into a fan of root-like fingers that match the ones on the Tulip Scroll vase exactly. The interior pattern is Zig-Zag and is found in the usual Millersburg colors.

FAN-TAIL

Fan-Tail is a pattern found on occasion as the interior design of Butterfly and Berry bowls. Actually, the design is made up of a series of peacock's tails swirling out from the center of the bowl, and while it is an interesting pattern, certainly is not a designer's success. This Fenton pattern has been reported in marigold, cobalt blue, green and white.

FANCIFUL

Probably many collectors will challenge this pattern being attributed to Dugan, but shards from the Dugan factory were identified by this author in both marigold and purple, so here it is. Actually, if a comparison is carefully made with the Embroidered Mums, Heart and Flowers and Fancy patterns, one finds a close similarity that isn't evident on first glance. Fanciful is available on bowls and plates in colors of marigold, peach, purple and white. The familiar Big Basketweave adorns the exterior.

Engraved Zinnia

Engraved Grape

Enameled Prism Band

English Hob and Button

Estate Mug

Fan

Evelyn

Fan-Star (Millersburg)

Fan-Tail

Fanciful

FANCY CUT

Much like the small creamers in design, Fancy Cut is actually the pitcher from a child's water set (tumblers to match are known in crystal). The color is a good rich marigold and the pitcher measures 4" tall and has a base diameter of 2¼". The design is a bit busy, combining several near-cut themes, but the rarity of the iridized version more than makes up for this.

FANS

The most striking thing about this pitcher is its size; larger than a creamer but smaller than a milk pitcher. It stands 5" tall and 7" across the handle. The design is a series of prisms in fan shapes and the only reported color is marigold. I suspect Davisons is the maker.

FARMYARD

Most people have long felt this to be Harry Northwood's masterpiece, but with the discovery of peach opalescent and evidence that the exterior pattern (Jeweled Heart) is Dugan, Farmyard is certainly from that company. Colors are purple, fiery amethyst, green and peach opalescent.

FASHION

Fashion is probably the most familiar geometric pattern in all of Carnival glass. It was manufactured in huge amounts over a long period of time and was originally called "402½" when first issued. The shapes are creamers, sugars, punch sets, water sets, a bride's basket, and a fruit bowl and stand. While marigold is the most often seen color, smoke, green and purple are found but are scarce. This pattern was made by Imperial. A rare red punch set was made since cups exist.

FEATHER STITCH

This pattern is a kissing cousin of the well-known Coin Dot design and as such is found on bowl shapes only, although I suspect a plate shape does exist. The colors seen are marigold, blue, green and amethyst and the bowls may vary in size from 8½" to 10". Feather Stitch was produced by Fenton.

FEATHER AND HEART

This fine Millersburg water set, found in green, marigold and amethyst is typical in many ways of all the Millersburg water sets. The pitcher lip is long and pulled; quite low while the rim is scalloped quite like its cousin, the Marilyn set. The glass is quite clear, rather heavy and has excellent iridescence. A little difficult to find, the pattern adds greatly to any collection.

FEATHERED FLOWERS

This pretty exterior pattern is found on the Australian Kiwi bowls. It is an intaglio pattern of swirls and stylized blossoms and is quite attractive.

Fancy Cut

Fans

Farmyard

Fashion

Feather Stitch

Feather and Heart

Feathered Flowers

FEATHERED SERPENT

At one time, this pattern was felt to be a Millersburg design, but in recent years the Fenton company has emerged as the known manufacturer. Found only in berry sets and a very rare spittoon whimsey, Feathered Serpent is available in marigold, green, blue and amethyst and has the Honeycomb and Clover as an exterior pattern.

FEATHERS

Northwood certainly made its share of vases perhaps because they were so decorative and useful. The Feathers vase is an average example usually found in marigold, purple and green and is of average size and quality.

FENTONIA

Known in berry sets, table sets, fruit bowls and water sets, Fentonia is an interesting all-over pattern of diamonds filled with the usual Fenton fillers of scales and embroidery stitchery. Colors are marigold, blue, green and occasionally amethyst, but the pattern is scarce in all colors and is seldom found for sale. There is a variant called Fentonia Fruit.

FERN

Again we find a pattern that is sometimes combined with the well-known Northwood pattern, Daisy and Plume. Fern is usually found as an interior pattern on bowls and compotes. It is an attractive pattern but really not an outstanding one.

FERN BRAND CHOCOLATES

Like so may of the small Northwood advertising pieces, this is a treasure for collectors and a real beauty for advertising buffs. The only color is amethyst, and it measures 6¼" across.

FERN PANELS

Like so many of the Fenton novelty pieces of Carnival glass, Fern Panels is found only on the hat shape. Not too exciting, but then it apparently had an appeal of its own then as now. The colors are the usual ones: marigold, blue, green and occasionally red.

FIELD FLOWER

Originally called "494½," this much overlooked Imperial pattern is really a little jewel. Found only on standard size water set and a rare milk pitcher, Field Flowers is found in marigold, purple, green and a very beautiful clambroth. The design, basically a flower framed by two strands of wheat on a stippled background, is bordered by double arches that edge the stippled area and continue down to divide the area into panels. All in all, this is a beautiful pattern that would grace any collection.

Fern Brand Chocolates

74

Feathered Serpent

Feathers

Fentonia

Fern

Fern Panels

Field Flower

FIELD THISTLE

Field Thistle is a U.S. Glass Company pattern, scarce in all shapes and colors. Shapes known are berry sets, table sets, water sets, a vase, plate and small creamer and sugar called a breakfast set. Colors most found are marigold or green, but the breakfast set is known in a beautiful ice blue so other colors may exist. The pattern is all intaglio and sometimes the marigold coloring is a bit weak.

FILE

For many years, File was designated as a product of the Columbia Glass Company, but was actually made by the Imperial Glass Company and is shown as such in their old catalogs in many shapes, including bowls, water pitchers, compotes and table sets. The colors are marigold, green, smoke and amethyst, and usually the luster is quite good with much gold in evidence. The mold work is far above average.

FINE CUT AND ROSES

The real pleasure of this Northwood pattern is the successful combination of a realistic floral pattern with a pleasing geometric one. Of course, rose bowls have a charm all their own. Really well done mold work and super color all add to its attraction, and even though it is slightly smaller than many rose bowl patterns, it is a favorite of collectors.

FINE CUT HEART (MILLERSBURG)

Here is the back pattern of the beautiful Primrose bowl, but the one shown is something special. An experimental piece with a "goofus" finish under iridization. I've seen one other Millersburg example of this on a Peacock and Urn ice cream bowl, and it was also amethyst.

FINE CUT OVALS (MILLERSBURG)

As a companion to the Whirling Leaves pattern, Fine Cut Ovals has eight ovals, each containing two sections of fine cut and two of file. At the edges, a fan flairs from a diamond center, giving interesting contrast.

FINE PRISMS AND DIAMONDS

This large English vase stands some 13½" tall and has a hefty base diameter of nearly 4". Obviously it was intended to be used and not just a decorative item. And while the design qualities aren't impressive, the glass is of good color and luster.

FINE RIB (NORTHWOOD)

Can you imagine anything simpler? Yet the Fine Rib pattern was and is a success to the extent it was used time and time again by the Northwood company as secondary patterns and is the primary one in attractive vases like the one shown. While common in marigold and purple, the green color is a scarce one in Fine Rib.

FINE RIB VASE (FENTON)

While both Northwood and Fenton produced a Fine Rib vase pattern, there are differences in the design. On the Northwood vase, the ribbing extends all the way down on the base, while the Fenton design ends in a distinctive scalloped edge above the base. Of course, Fenton made this shape and pattern in red too.

Fine Cut Heart (Millersburg)

Fine Cut Ovals (Millersburg)

Field Thistle

File

Fine Cut and Roses

Fine Prisms and Diamonds

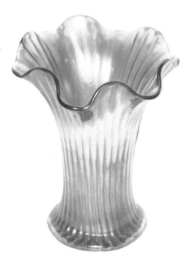

Fine Rib

Fine Rib Vase

FINECUT RINGS

From a set of copyright drawings we know this pattern was made by the Guggenheim, Ltd. Company of London in 1925. Shapes shown in the drawing, are an oval bowl, vase, footed celery, covered butter dish, creamer, stemmed sugar, round bowl, footed cake plate and covered jam jar. The only color I've seen is marigold and as you can see, the coloring, mold work and luster are outstanding.

FISH VASE

This handsome vase has finally been traced to the makers of the Zipper Stitch pattern as well as the Hand Vase and is believed to come from Czechoslovakia. I've seen only marigold, but have had both amethyst and green reported to me by a British collector.

FISHERMAN'S MUG

While I'm not completely convinced this is a Dugan pattern, it seems to be by a process of elimination. At any rate, it is much sought, and like the more rare Heron mug, has the pattern on one side only. Colors are purple, amethyst, marigold, a rare peach opalescent and a more rare blue.

FIVE HEARTS

Like so many of the Dugan patterns in peach opalescent, Five Hearts is found only on dome-footed bowls of average size. It has been estimated that 60% to 75% of all peach Carnival was made by this company. Personally, I'd say it would be closer to 90%.

FLARED WIDE PANEL ATOMIZER

I've often wondered just how many iridized patterns in atomizers there really are. Certainly it was a popular item in the days of Carnival, and each of them seem to have a charm of its own. This one, while rather plain, is a real cutie.

FLEUR DE LIS (MILLERSBURG)

Named, of course, for the stylized figures which symbolize the lily of the French royal family, this beautiful Millersburg design is found on bowls of all shapes and as an interior pattern on an occasional Hobstar and Feather punch bowl. The overall pattern is formal though well-balanced, and the quality of workmanship, iridescence and color rank with the best. Especially beautiful is the dome-footed three-cornered bowl in amethyst.

FLEUR DE LIS VASE (JENKINS)

Like other Jenkins patterns, this heavy vase has a deeply intaglio pattern. It stands a stately 10½" tall and is found on a rich marigold.

Flared Wide Panel Atomizer

Finecut Rings

Fish Vase

Fisherman's Mug

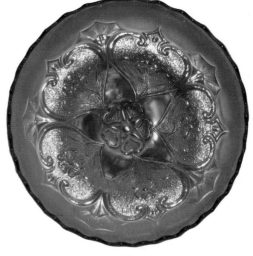

Five Hearts

Fleur De Lis

Fleur De Lis Vase

FLORAL AND GRAPE

Without question, Floral and Grape is one of the most familiar Fenton patterns. The water sets are plentiful, especially in marigold. The only other shape is a hat whimsey pulled from the tumbler and other colors are cobalt blue, amethyst, green and a scarce white. Also, there are variations due to the wear of molds and creation of new ones. A variant is credited to the Dugan company.

FLORAL AND OPTIC

Most people are familiar with this pattern in rather large footed bowls in marigold or clear Carnival, although it exists on rare cake plates and rose bowls, and in smoke, iridized milk glass, red, a stunning aqua, white and iridized custard glass. The pattern is quite simple, a series of wide panels edged by a border band of vining flower and leaf design. A green or amethyst bowl in Floral and Optic would be a real find in this Imperial pattern.

FLORAL AND WHEAT

Found in both stemmed bonbons and compotes, this pattern is said to come from U.S. Glass and is one of the more often found patterns. It can be found in marigold, amethyst, and blue as well as peach opalescent and the coloring is usually quite rich.

FLORENTINE

Both Fenton and Northwood made these candlesticks, and it is difficult to tell them apart. These shown are the Fenton version, however. They come in many colors including pastels and red as well as the usual Carnival glass colors.

FLOWER BLOCK

This pattern is a spin-off of the Curved Star pattern, and I'm showing it here in its entirety so collectors will be able to see the complete block as designed. Most of the ones around have lost their fancy wire base and wire arranger on the top.

FLOWER POT (BLUE)

While both Imperial and Fenton made these flower pot and saucer planters, the one shown is from the Fenton Art Glass Company. It is a beautiful ice blue and is 5" tall and 4¾" in diameter. The plate is 6¼" in diameter. It has been seen mostly on pastels including pink but can also be found on marigold.

FLOWERING DILL

Once again we encounter a Fenton pattern that was chosen for the hat shape only, but Flowering Dill has a bit more to offer than some in that the design is graceful, flowing and covers much of the allowed space. Flowering Dill can be found in marigold, cobalt blue, green and red.

FLOWERING VINE

The Flowering Vine compote is indeed a very scarce Millersburg item. To date, I've heard of only two examples, one in green and one in amethyst. The compote is a large one, some 9" high and 6½" across. The interior pattern is one of grape-like leaves and a dahlia-like flower. The finish is a fine radium one on heavy glass.

Floral And Wheat

Florentine

Floral and Grape

Floral and Optic

Flower Block

Flower Pot

Flowering Vine

Flowering Dill

FLOWERS AND FRAMES

Here is another of Dugan's dome-footed bowls found primarily in peach opalescent, but also available in marigold as well as purple and green. The bowls vary from 8" to 10" depending on the crimping of the rim and usually have very sharp mold detail.

FLUFFY PEACOCK

Once thought to be a Millersburg product (it's really that pretty), Fluffy Peacock is now recognized as a Fenton product, and as such is high on the list of their best in iridized glass. Known only in water sets, Fluffy Peacock is not too difficult to find in green or marigold; however, the beautiful cobalt blue is quite scarce, and the amethyst is much sought.

MILLERSBURG FLUTE

The Millersburg Flute has one or two distinct differences from any of the other Flute patterns. One, of course, is the clarity of the base glass itself but the most pronounced is the ending of the flutes themselves plus the clover leaf base. It is found in punch sets, berry sets, vases and a rare compote.

FLUTE (NORTHWOOD)

While it is certainly true that all of the major makers of Carnival glass used the Flute pattern one way or another, apparently only Imperial and Northwood thought enough of it to make it a primary pattern. Thus we find many useful shapes in an array of colors coming from the Northwood factories – including sherbets, water sets, table sets, berry sets and even individual salt dips (an item seldom encountered in Carnival glass). The green water set is probably the rarest color and shape in this pattern.

FLUTE #3 (IMPERIAL)

After I wrote *Millersburg, Queen of Carnival Glass,* a great controversy arose around the Flute pattern showed in a small bowl and punch set. Many people thought this was the Imperial pattern called Flute #3 (it wasn't), which I've seen only on table sets, water sets, a celery or large spooner, a small toothpick holder, a sauce bowl, a two-handled toothpick holder and a different punch set (since then I've seen a very large berry bowl). The obvious difference between these two Flutes is the bottom configuration. On the Imperial version, there is a definite hexagonal effect just above the base, while the Millersburg version is etiher slightly curved outward in a honeycomb effect or has a a curved slope from the base upward.

FLUTE AND CANE (IMPERIAL)

Found in wines, champagnes, punch cups, a milk pitcher and the water set pieces shown, this is a very scarce pattern indeed. All shapes are found in marigold, but the water pitcher is reported in white also and is much sought in either color as are the marigold tumblers.

FLUTE AND HONEYCOMB (MILLERSBURG)

Here's a better look at the 5" bowl I showed in *Millersburg, Queen of Carnival Glass.* You can see by the design above the base why I've renamed it rather than simply calling it a Flute variant. Super amethyst is the only color I've seen, and it matches the Big Thistle punch bowl in color and finish.

FLUTE SHERBET (ENGLISH)

This little cutie is signed "British" in script on the underside of the base. In size they are small, measuring 3¼" tall and 3" across the top. The only color reported is a good marigold of deep hue.

Flute and Cane (Imperial)

Flowers and Frames

Fluffy Peacock

Millersburg Flute

Flute

Flute #3

Flute and Honeycomb

Flute Sherbet

FLUTED SCROLLS

Perhaps this should be called a spittoon, but regardless of the name if is a very rare, one-of-a-kind produc in Carnival. While often found on opalescent glass, this is the only iridized item in this pattern I've heard of Perhaps it was a novelty a Dugan worker produced for himself. The coloring is a good amethyst with averag luster.

FOLDING FAN COMPOTE

The only shape I've seen for this Dugan pattern is the stemmed shallow compote shown. Most have the ruffle edge, and the average size is 7½" in diameter and 4" tall. Colors are marigold, amethyst, green and peach opalescent. The one shown has a clear base, but this isn't always the case.

FOOTED PRISM PANELS

What a nice design this pretty 10" footed vase is! It is a Sowerby product, and the only reported colors are a good rich marigold and a scarce green. The design is a series of six panels filled with star prisms in graduating sizes. The stem base is domed and gently scalloped, giving great space to the appearance.

FOOTED SHELL

Made in two sizes, 5" and 3", this little novelty from the Westmoreland company is a scarce item, especially in the smaller size. The shell rests on three stubby feet and the iridization is on the inside only. Colors I've seen are amethyst, green, marigold and blue. Also milk glass iridized.

FORKS

I first showed this rare carcker jar in my *Rarities in Carnival Glass* book as an unlisted pattern but have since learned it was called "Forks" in old Cambridge ads. The only color I've seen is a very rich green, but I wouldn' rule out marigold or amethyst.

FORMAL

I'm a wee bit skeptical about this pattern, but new evidence points to Dugan as the maker. Nevertheless, i is an interesting pattern found only on the vase shaped in a jack-in-the-pulpit manner and a quite scarce hatpir holder. The colors most seen are purple and marigold, but I'm told pastels do exist. The hatpin holder stand: 7¼" tall.

FOUR FLOWERS

For many years I've been puzzled by the origin of this beautiful design, but since peach opalescent is a prominent color, I'm satisfied it was also a Dugan product. Known in bowls and plates from 6" to 11", colors are purple, green, marigold, peach, blue and smoke.

FOUR FLOWERS VARIANT

With all the recent research on this pattern, I seem to grow even more confused; but for argument's sake, I'l say the theory now seems to place this pattern's maker in the Scandanavian orbit of glass making. It is a beauty whoever made it, much prettier than the regular Four Flowers and is found in bowls, plates and a stunning salver on a metal base.

Four Flowers

Folding Fan Compote

Fluted Scrolls Footed Rose Bowl

Footed Prism Panels

Footed Shell

Forks

Formal

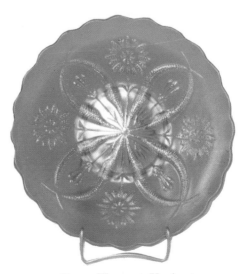

Four Flowers Variant

474

Here is a very pretty near-cut pattern that really looks quite good on all shapes. I known of punch sets, table sets, water sets and a milk pitcher in 474, as well as a nice fluted bowl. The color is usually marigold but amethyst and green do exist in some shapes and are highly prized. The glass is heavy and clear, the coloring extra fine and the mold work superior; all making Imperial's 474 a real treat to own. Rare vases are also known.

FRENCH KNOTS

I suppose most people credit this pattern to Fenton because of the shape – a typical hat with ruffled top. The design is quite nice, however, and gracefully covers most of the space except the base. The exterior is plain but nicely iridized, and the only colors I've heard about are marigold and blue. French Knots is 4" tall and has a base diameter of 2½".

FROLICKING BEARS

Rare is hardly the word for this distinguished pattern, and I certainly wish I could identify the maker positively, but I can't (although I lean toward the U.S. Glass Company). The coloring is an odd gunmetal luster over an olive green glass. The mold work is good but not exceptional.

To date, the Frolicking Bears tumbler has sold for more money than any other, and the pitchers rank near the top of the market.

FROSTED BLOCK

Frosted Block is a very Imperial pattern found in several shapes including bowls, creamers, sugars, compotes, plates, milk pitchers, pickle dishes and rose bowls. The color most seen is a good even marigold, but as you can see from the photo, a beautiful fiery clambroth is known in most shapes. In addition, some pieces are marked "Made in USA." These pieces are much scarcer and are well worth looking for.

FRUIT AND FLOWERS

Apparently a very close relative to the Three Fruits pattern, this is another of Northwood's floral groupings, so well designed and produced. Also, there are several variations of this pattern, some with more flowers intermingled with the apples, pears and cherries, often meandering almost to the very outer edge of the glass. Again the Northwood Basketweave is the exterior pattern, and Fruits and Flowers is found on compotes with handles as well as bowls and rare plates.

MILLERSBURG FRUIT BASKET

I couldn't quite believe my eyes when I first saw this Millersburg compote. But there it was with the exterior design and shape exactly like the Roses and Fruit Compote with an interior basketweave design and a pineapple, grapes and fruit! It was a real find, and I've been grateful ever since for the privilege of photographing it. It is my belief that this compote was the original pattern design but had to be modified because of the difficulty in producing the intricate design and remain in a competitive price area. At this time, four have been found. Of course, it must be classified as an extreme rarity.

FRUIT SALAD

Westmoreland seems to be the maker of this beautiful and rare punch set after years of speculating. Found in marigold, amethyst and peach opal, the design is outstanding and desirable.

Fruit Salad

474

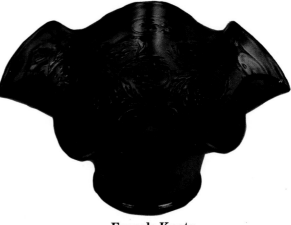

French Knots

Frolicking Bears

Frosted Block

Fruits and Flowers

Millersburg Fruit Basket

GARDEN MUMS

Here is the Northwood design used on so many of their advertising pieces. The flat plate is 6" across, and the coloring is a typical amethyst. Just why these few examples were left unlettered is a mystery, but they certainly add much to any collection.

GARDEN PATH

Both the regular and the variant of this pattern are outstanding examples of pattern work, and the 11" plate shown points this out. Colors are marigold, amethyst, peach opal and white for the regular Garden Path, and amethyst and peach for the variant.

GARLAND

Quite often you'll spot one of these nicely done footed rose bowls in cobalt blue or marigold, but the green is quite rare, and I haven't heard of an amethyst one, but I suspect it was made in that color. Made around 1911 these are shown in Fenton ads well up into the later dates of iridized glass and must have been quite popular. There are three sets of wreaths and drapery around the heavily stippled bowl.

GAY 90'S

Not only because of its extreme rarity but because the design is so very well suited for a water set is this pattern recognized as one of two or three top water sets in all of Carnival glass. In every regard, excellence of workmanship is obvious. Even the solid glass handle shows leaf veining at the top and intricate petal sliping at the base. Add to this the beautiful Millersburg finish and the clarity of superior glass, and you have a real winner – the Gay 90's water set.

GEORGE W. GETTS PLATE

Here is still another of the Northwood advertising small plates with the Garden Mums design. As you can see this one's claim to fame is the backward "S" in "Grand Rapids." Whether intentional or not, it catches the eye. Amethyst is the only reported color.

GIBSON GIRL

What a cute little toothpick holder this is despite having no pattern whatsoever! It stands 2⅜" tall. Who made it and in what other colors I haven't a clue.

GOD AND HOME (DUGAN)

For many years, little was known about this unusual water set, except that it was found only in cobalt blue. It has a laurel wreath, a sun and rays. On one side it reads "In God We Trust," and on the other side, "God Bless Our Home." This rare set is now thought to have been made at the Dugan/Diamond factory, and it is a real pleasure to show it here, thanks to the Yohes of Stuttgart.

GODDESS OF HARVEST

Goddess of Harvest is the rarest of all Fenton bowl patterns and certainly deserves all the attention it gets. I've heard of six or seven of these beauties in colors of marigold, blue and amethyst and each is highly treasured by its owner. The bowl measures about 9" in diameter and is usually found with a candy ribbon edge.

GOLDEN CUPID

This very scarce Australian beauty is something to behold. In size it is only 5¼" in diameter and quite shallow. The glass is clear with the cupid in guilt and a strong iridescence over the surface. It may be an ashtray, but I can't be sure. Large bowls are reported also.

George W. Getts Plate

Garden Mums

God and Home (Dugan)

Garden Path

Garland

Gibson Girl

Gay 90's

Goddess of Harvest

Golden Cupid

GOLDEN GRAPE

Known only in bowls or rose bowls on a collar base, this is a neatly molded item without much elaboration. The exterior is completely plain, and the only colors found are green, marigold or pastel marigold, usually with a satin finish. Golden Grape was manufactured by Dugan.

GOLDEN HARVEST

Here is another U.S. Glass pattern found mostly in marigold but occasionally seen in amethyst and reported in white. As you will notice, the wines are different from the decanter but are the proper ones, and the stopper is solid glass.

GOLDEN HONEYCOMB

This interesting Imperial pattern provides an all-over design without being busy. The small bowls have odd little solid glass handles that are like the ones on the breakfast set while the plate and compote do not. The only color I've heard of is a good deep marigold with very rich iridescence. While this certainly isn't in the same class as the Dugan Honeycomb rose bowl, it is a better than average item to own.

GOLDEN OXEN

While this mug has never been one of by favorites like the Heron, it is hard enough to find so that many people want one. Taller than most mugs, the only color I've seen in the marigold.

GOLDEN WEDDING

I am very pleased to show the complete item, just as sold. Notice that the cap is still sealed, the whiskey is still inside, the labels are intact and the beautiful box, dated December 31, 1924, gives us a time frame for the bottle. Several sizes exist from a full quart down to the tiny $\frac{1}{10}$ pint size.

GOOD LUCK

I'm very sure every Carnival Glass collector is familiar with this pattern since most of us have found a place for an example in our collections at one time or another. The example shown is quite unusual in that it is a true aqua blue with reddish iridescence. It is the only example I've ever seen in this exact color although the Good Luck bowls and plates are found in a very wide range of colors. This piece was made by Northwood.

GOTHIC ARCHES

Unlisted up until now, this beautiful vase stands 10" tall. The flared or morning glory style top adds to the interest as does the fine smoke coloring. The maker is unknown, but I suspect Imperial. To date other colors have not been reported but probably exist.

GRACEFUL

Once again we find a very simple pattern, so very different from most Northwood offerings in the Carnival glass field. Found mostly in marigold, occasionally a rich purple or deep emerald green vase in this pattern will surface and when one of these is found, the simple beauty of the Graceful pattern becomes obvious.

Gothic Arches

Golden Oxen

Golden Grape

Golden Harvest

Golden Honeycomb

Golden Wedding

Good Luck

Graceful

91

GRAND THISTLE (PANELED THISTLE)

Regardless of which of the three names you choose to call this pattern (the third is Alexander Floral), the rarity and the coloring demand attention. It was made at one (or more) of the glass factories in Finland and only in blue.

GRAPE (IMPERIAL)

Perhaps reproductions have detracted too much from this beautiful Imperial pattern for most of us, but regardless of that, it remains a beautifully designed, nicely done pattern. In fact, for sheer realism, it doesn't take a back seat to any Grape pattern! The shapes are almost endless, and the colors range from marigold to amethyst, green, smoke, clambroth and amber, all usually with a very fine luster.

GRAPE AND CABLE (FENTON)

Yes, Fenton made a Grape and Cable pattern, and it is often hard to distinguish it from Northwood's. Fenton's contribution is found in bowls (both flat and footed), plates and large orange bowls usually with the Persian Medallion interior. Colors are marigold, green, amethyst, blue and rarely red.

GRAPE (AND CABLE) (NORTHWOOD)

Northwood Grape is, without question, the all-time favorite in Carnival glass. Not only did it lead the field 60 years ago, it still does today. The variety of shapes available is staggering, and the color availability large; the pastels are especially sought. Of course, this pattern usually brings top dollars, especially in the rare shapes and colors. A small spittoon sold for $7,000.00 at the Wishard auction. Whatever the undeniable appeal, this uncomplicated pattern has to be rated as the top item and entire collections of the Grape aren't uncommon.

GRAPE AND CHERRY

Known only on large bowls, this Sowerby pattern is a real beauty. The design is exterior and all intaglio and is a series of grapes and cherries separated by an unusual torch and scroll design. The base has a grape and leaf design, also intaglio. The only colors I've heard about are marigold and cobalt blue, but others may certainly exist.

GRAPE AND GOTHIC ARCHES

Made in a variety of kinds of glass including crystal, custard, gold decorated, and Carnvial glass, this pattern is certainly one of the earlier grape patterns. The arches are very effective, reminding one of a lacy arbor framing the grapes and leaves. While the berry sets often go unnoticed, the water sets are very desirable and are a must for all Northwood collectors.

GRAPE ARBOR

This is an underrated pattern, especially in the large footed bowl shape which carries the same exterior pattern as the Butterfly and Tulip. Of course, the tankard water set is popular, especially in the pastel colors of ice blue, ice green and white. The marigold set seldom brings top dollar, and this is a shame since it is quite nice. The only other shape in Grape Arbor is a scarce hat shape.

Grand Thistle

Grape (Imperial)

Grape and Cable (Fenton)

Grape (and Cable), (Northwood)

Grape and Cherry

Grape and Gothic Arches

Grape Arbor

GRAPE DELIGHT

Here is a pattern I'm sure will bring on a few outcries, because I've often heard it declared to be a Fento product. I'm very sure, however, that it came from the Dugan family. On close comparison with several Duga products, the mold work is certainly compatible. Not only does it come in the scarce nut bowl shape shown bu in the more often seen rose bowl. The colors are both vivid and pastel and the most unusual feature, the six stuff feet.

GRAPE LEAVES (MILLERSBURG)

Seldom found and always costly, this rare Millersburg pattern is quite like the Blackberry, Grape an Strawberry Wreath patterns except for its rarity. The major design difference lies in the center leaf and grapes and, it is found in marigold, amethyst, green and vaseline.

GRAPE LEAVES (NORTHWOOD)

Harry Northwood must have loved this pattern, for I've never seen a poor example. The purple bowls are almost without exception, vividly brilliant with strong color and iridescence. The ice blue is one of the pretties items in this color I've seen and green has also been found. It is sad no other shape was chosen for the Grap Leaves pattern, but this is the case. The exterior is the familiar Wild Rose pattern with a finely stipple background.

GRAPE WREATH

This bowl might be called the "missing link" for it stands squarely between the Blackberry Wreath and th Millersburg Strawberry and seems to be a part of the series – perhaps from the same designer.

Besides the various sizes of bowls, a rare spittoon whimsey, 6" and 10" plates are known.

GRAPE WREATH VARIANT (MILLERSBURG)

Actually, only the center design has been changed on these and as you can see, it is a stylized sunburst. I'v seen several sizes of bowls with this center but it looks best on the large 10" ice cream bowls. Colors are marigold green, amethyst and a beautiful clambroth with much fire in the luster.

GRAPEVINE LATTICE

I personally doubt the plate is really the same pattern as the water set known by the same name but wil give in to tradition and list them as one and the same. The plate and bowl could be called "Twigs" since the closely resemble the Apple Blossom Twigs pattern minus the flowers and leaves. The colors are both vivid an pastels, usually with very good iridescence. This pattern was manufactured by Dugan.

GREEK KEY

This simple continuous Northwood pattern, called Roman Key in pressed glass, is really very attractive whe iridized. Acutally, there are three motifs used: the Ray center pattern, the Greek Key and the Beads (in the muc sought water set a fourth motif of prisms is added). Besides the water set, there are flat and footed bowls as we as plates. The colors are both vivid and pastels.

Grape Wreath Variant

Grape Delight

**Grape Leaves
(Millersburg)**

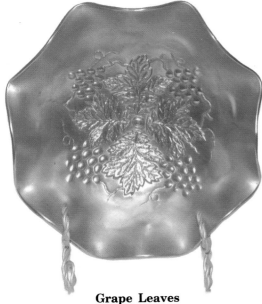

**Grape Leaves
(Northwood)**

Grape Wreath

Grapevine Lattice

Greek Key

95

GREEK KEY AND SCALES (NORTHWOOD)

Here's a rather nice Northwood secondary pattern usually found on dome-based bowls with the Stippled Ray variant interior. The design is crisp and does a nice job.

HAMMERED BELL

Whoever made this scarce and attractive light shade should be most proud for it is an imaginative work of art. Found only on a frosty white, the pattern is clear and distinct. The metal handle may vary in design, but all were used to suspend the bell **above** or **below** the bulb.

HANDLED VASE (IMPERIAL)

Mrs. Hartung shows this neat vase with the lip turned down, and Rose Presznick drew it cupped in like a rose bowl. But here it is in the original shape, and it is very pretty indeed. It stands about 9" tall, and the coloring is a deep rich marigold well down before paling to a lighter shade.

HARVEST FLOWER

Here is a seldom-seen pattern, found only on a pitcher and tumbler in marigold, amethyst and green on the latter and only in marigold in the pitcher. The tumblers have been reproduced in a dingy purple as well as white so beware.

HATTIE

While there is little to be called outstanding about Hattie, it nonetheless has its own charm and has the distinction of having the same pattern on both the exterior and interior. Most often found on 8" bowls with a collar base, it is also found in a scarce rose bowl and two sizes of plates, both of which are rare. The color most seen on this Imperial product is marigold, but green and amethyst do exist, and I've heard of both amber and smoke bowls.

HEADDRESS

Here is a pattern found in two varieties because it was made both in America and in Sweden. It is found in 9" bowls and on compotes in marigold, green and blue.

HEART AND HORSESHOE

Yes, Fenton had a version of the Good Luck pattern, and it is much harder to find than the one Northwood produced. Note that it is simply the familiar Heart and Vine pattern with the horseshoe and lettering added. Colors are marigold and green with the marigold most prevalent and the green very hard to find.

Greek Key and Scales

Hammered Bell

Harvest Flower

Hattie

Handled Vase

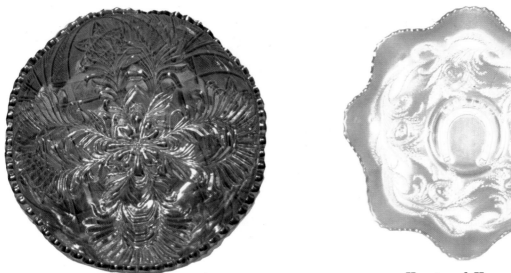

Headdress

Heart and Horseshoe

HEART AND TREES

Found as an interior pattern on some of the footed Butterfly and Berry bowls, Heart and Trees is combination of three well-known Fenton standards. Colors known are marigold, green and blue but certain others may exist.

HEART AND VINE

Apparently this was one of Mr. Fenton's favorite designs for he used several variations of it combined with the Butterfly and Berry pattern as Stream of Hearts or on the Heart and Horseshoe pattern. The Heart and Vine pattern can be found on both bowls and plates and occasionally with advertising on the latter (the Spector plate Colors are marigold, green, amethyst, blue and white.

HEARTS AND FLOWERS

Hearts and Flowers is an intricate pattern, like so many others favored by the Northwood company. It is found in various sizes and shapes of bowls as well as small, very graceful compotes. Apparently it was designed as close relative of the Fanciful and Embroidered Mums patterns, perhaps by the same moldmaker. Interestingly enough, the compotes in the pattern are very similar in shape to the Persian Medallion compotes attributed to the Fenton Glass Company. Shown is a rare plate in sapphire blue.

HEAVY BANDED DIAMONDS

Perhaps designed as a companion to the Banded Diamonds water set, this beautiful Australian berry set available in both marigold and purple, is a joy to behold. The diamonds are heavily molded and stand out below the narrow thread lines and the iridescence is very rich.

HEAVY DIAMOND

I'd always thought this pattern to be a Dugan product until this vase shape in smoke appeared, so apparently it is an Imperial product. Previously reported only in marigold, shapes known besides the vases are large bowls creamers and sugars, but others probably are around.

HEAVY GRAPE (DUGAN)

So similar in many ways to the Millersburg Vintage bowls, this Dugan pattern does have several distinctive qualities of its own. The most obvious one is, of course, the grape leaf center with grapes around its edge. Also missing are the usual trendrils and the small leaflets, and the exterior doesn't carry a hobnail pattern but typical near-cut design.

HEAVY GRAPE (IMPERIAL)

Once called a Fenton product, this beautiful Grape design is now known to be an Imperial pattern and i available on berry sets, nappies, custard sets and plates in three sizes. The colors on most pieces are simply spectacular and include marigold, green, purple, smoke, amber, pastel green, smokey blue and clambroth. The 11 chop plate is highly sought and always brings top dollar when sold.

Heavy Grape (Imperial)

Heart and Trees

Heart and Vine

Heart and Flowers

Heavy Banded Diamonds

Heavy Diamond

Heavy Grape (Dugan)

HEAVY HOBNAIL

I first showed this unusual item in *Rarities in Carnival Glass*. At that time, only the white version was known but since then a spectacular purple example has been seen. These were the Fenton Rustic vases that weren't pulled or slung into the vase shape and very few are known.

HEAVY IRIS

I'm sure every collector of Carnival glass has speculated at one time or another over the maker of this beautiful pattern. Some have felt Millersburg, other Fenton. Now, however, it is with some assurance we report that a very sizeable chunk of a Heavy Iris tumbler was unearthed at the Dugan dump site in 1975 so we can declare it a Dugan pattern. The shape is, of course, a graceful tankard water set. The design is sharp and heavy, simple enough to be effective yet covering much of the available space. The colors are marigold, purple and white with purple the most desired. All in all, a beautiful set to own.

HEAVY PINEAPPLE

I feel very happy to show this Fenton rarity because it is only the second example in Carnival glass to be reported. This one is a beautiful frosty white while the other was cobalt blue. Both pieces are large 10" bowl on three feet. The design is all exterior and heavily raised. Certainly there must be other examples of Heavy Pineapple but if so, they haven't been reported. A third marigold example is known.

HEAVY PRISMS

This beautiful celery vase stands 6" tall and shows the quality English Carnival glass makers attained. The maker is Davisons, and the colors reported are marigold, amethyst and blue. The glass is very thick and heavy and the luster top notch.

HEAVY SHELL

Found only on white Carnival, this Dugan pattern is rather scarce. The shapes are oval bowls and matching candlesticks. The glass is quite heavy and the luster very rich.

HEAVY WEB

Found primarily on large, thick bowls in peach opalescent, Heavy Web is a very interesting Dugan pattern. It has been found with two distinct exterior patterns – a beautifully realistic grape and leaf design covering most of the surface and an equally attractive morning glory pattern. I've seen various shapes including round, ruffled, square and elongated ones. A vivid purple or green bowl in this pattern would certainly be a treasure.

Heavy Hobnail

Heavy Iris

Heavy Pineapple

Heavy Prisms

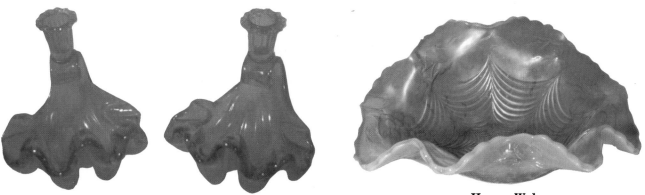

Heavy Shell

Heavy Web

HERON

Like the Fisherman mug shown elsewhere in this book, the Dugan attribution is sheer speculation and has been arrived at only by the process of elimination. They simply do not look like products of any of the other known companies that made sizeable amounts of iridized glass. Known only in amethyst, and a very rare marigold, these mugs are much harder to find and must be considered rare. The pattern is on one side of the mug only.

HOBNAIL

While plentiful in many other types of glass, Hobnail is quite a rare item in Carnival and one of real beauty. The very pattern seems perfectly suited for iridescence. The base carries a many-rayed design, and all the pieces I've seen are top-notch. The pitchers and tumblers are very rare and extremely hard to locate but even the rose bowl is a prize and the lady's spittoon is a little darling. Hobnail was made by Millersburg.

HOBNAIL VARIANT

This rare and exciting Millersburg pattern is quite different from the regular Hobnail made by the same company, in that it has 11 rows of dots with 18 in each row around. The top row has only nine hobs however. I suspect this was a copycat item from Fenton's Heavy Hobnail, but it is definitely Millersburg since mold drawings from there exist. Shapes known are vases, rose bowls and a jardiniere. Colors reported are marigold, amethyst and green.

HOBNAIL SODA GOLD

While most spittoons found in Carnival glass are of a daintier size, Imperial's Hobnail Soda Gold spittoon is a larger, more practical size. It measures 7" across the top and stands 5" high. I've seen examples in marigold and green as well a a peculiar dark shade of amber. Many of these have a great deal of wear on the bottom indicating they were actually used!

HOBSTAR

Apparently this was one of the early near-cut patterns from Imperial. I'd guess it experienced great popularity from the first for it was carried over from crystal to Carnival glass in many shapes, including berry sets, cookie jars, table sets, bride's baskets and a very rare pickle caster in an ornate holder. Marigold is the common color with purple and green quite scarce.

HOBSTAR AND ARCHES

Here is another well-done geometric design from the Imperial Company. Found mostly in bowls, a two-piece fruit bowl is available, as shown. Please note the base is the same as that of the Long Hobstar set. Both are correct, and it was unusual for a company to get double mileage where possible. Colors are marigold, green, amethyst and smoke.

HOBSTAR AND CUT TRIANGLES

This very unusual pattern is typically English in design, with strong contrast between geometrically patterned areas against very plain areas. Rose Presznick lists bowls, plate and a rose bowl in this pattern in both green and amethyst, but the only shapes I've seen are bowls, rose bowls and compotes in marigold or amethyst.

Hobnail Variant

102

Heron

Hobnail

Hobnail Soda Gold

Hobstar

Hobstar and Arches

Hobstar and Cut Triangles

HOBSTAR AND FEATHER

If one word could be summoned to describe this Millersburg pattern, that word would be "massive." The glass is thick, the near-cut design deeply cut and impressive. Whether it is a punch bowl, a compote, a giant rose bowl or a pulled vase whimsey, the Hobstar and Feather pattern is not likely to be confused with any other. The usual colors of marigold, green and amethyst are to be found and always in a fine radium finish.

HOBSTAR AND FILE

What a pleasure it is to be able to list this very rare water pitcher (the only one known). Besides the pitcher two tumblers have been reported. The coloring is a good marigold, and the mold work excellent. It is probably Imperial, but I have no confirmation at this time. (Photograph not available)

HOBSTAR AND FRUIT

This Westmoreland pattern is known on small and large bowls and a rare 10½" plate. The bowls are found in peach opalescent mostly, but the small bowl is known in aqua opalescent, and the plate has been found in ice blue.

HOBSTAR BAND

This scarce Imperial pattern is found only on handled celery vases and water sets of two varieties. The usual pitcher is flat based, but as you can see a quite rare pedestal variant is known. The only color reported is a good rich marigold.

HOBSTAR FLOWER

This beautiful little Northwood compote is seldom seen and usually comes as a surprise to most collectors. I've heard it called "Octagon" or "Fashion" at one time or another, and most people say they've never seen it before. It is a rather hard-to-find item. It is known in marigold, but is mostly found in amethyst.

HOBSTAR PANELS

Apparently the English glassmakers like the creamer shape, for they certainly made their share of them. Here is a well-conceived geometric design with hobstars, sunbursts and panels integrated into a very nice whole. The only color I've seen is the deep marigold shown, but certainly there may be others.

HOBSTAR REVERSED

Shown is what was called a "sideboard set" in the advertising by Davisons of Gateshead. It consists of a flower frog and holder that is footed and two side vases which were also used as spooners. Also known is a covered butter dish. The colors listed are marigold, amethyst and blue, but not all colors are found in all shapes.

Hobstar and Feather

Hobstar and Fruit

Hobstar Band

Hobstar Flower

Hobstar Panels

Hobstar Reversed

HOBSTAR WHIRL (WHIRLIGIG)

While I can't be certain who made this very nice compote, I wouldn't rule out Dugan or Northwood. The design is simple but effective, and the luster very good. The coloring is cobalt blue.

HOLIDAY

Holiday is a Northwood pattern that so far has turned up only on this 11" tray. It is shown here in marigold and has the Northwood mark. It is also known in crystal but seems to have no accompanying pieces.

HOLLY

Holly is one of the more common Fenton patterns being available in bowls, compotes, goblets, hat shapes and plates. The design is a sound one and tends to look good, especially on the larger bowl and plate shapes. Colors are marigold, green, amethyst, fiery amethyst, blue, aqua opal, white vaseline, ice blue and red. Shown is a beautiful marigold opalescent.

HOLLY AND BERRY

This Dugan pattern, found mostly in rather deep bowls and nappies (one-handled bonbons) as well as an occasional sauce boat, is so similar to the Millersburg Holly pattern that most people simply accept it as a product of that company. The one distinguishing difference is the leaf and berry medallion in the center of the Holly and Berry pattern while the Millersburg design has no such center motif. Holly and Berry is found in purple, green, marigold, blue and most often peach opalescent. A large piece of this pattern was discovered in the Dugan diggings.

HOLLY SPRIG (or WHIRL)

I have combined these two Millersburg patterns under one title because, frankly, I can distinguish no discernible difference in them. Either can be found with or without a wide panel or the near-cut wreath pattern. The holly leaves and berries do vary somewhat from shape to shape but certainly not enough to warrant separate titles. At any rate, it is a simple, well-executed pattern, especially nice on a one-handled nappy. The colors are usually green, marigold and a fine amethyst.

HOLLY WREATH VARIANT (MILLERSBURG)

Besides the Grape Wreath variants, Millersburg's Holly patterns were often given the three same center patterns of feathers, stylized clovers and multi-ringed stars. Here is a very pretty example of the feather center on a bowl with a candy-ribbon edge. All of these are found on 7" – 8" bowls in the usual Millersburg colors.

HONEYBEE POT

First drawn by Rose Presznick, this cute novelty item is first rate. The coloring is very yellow and really suits the purpose in every way with even a slot for a spoon to dip the honey.

HONEYCOMB AND CLOVER

While this Fenton pattern is best known as the exterior design of Feathered Serpent pieces, it is also used as the primary pattern for bonbons and the scarce compote. Colors I've heard about include marigold, green, blue, amethyst, amber and white but others may exist!

Honeybee Pot

Holly Wreath Variant
(Millersburg)

Hobstar Whirl

Holiday

Holly

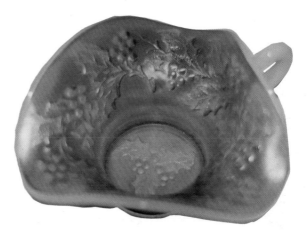

Holly and Berry

Holly Sprig (or Whirl)

Honeycomb and Clover

HONEYCOMB AND HOBSTAR

This very rare vase is from the Millersburg company and has never been out of the Ohio area. As you can see, it is a very rich blue with heavy even luster. The base design is the same as the Hobstar and Feather rose bowl. The vase stands 8¼" tall. This example was found years ago on a shelf in a defunct Millersburg business and has remained in its present home for many years. Another blue and one amethyst are reported.

HONEYCOMB ROSE BOWL

The latest shards from the Dugan digs have tuned up two large pieces of this pattern, so Dugan was the maker. The Honeycomb Rose bowl is a rare item, found in both marigold and peach opalescent.

HORN OF PLENTY

What can I say about this bottle that isn't obvious from the photo? It isn't rare or even good Carnival, but it certainly has a place in the history of our glass.

HORSES' HEADS

Sometimes called Horse Medallion, this well-known Fenton pattern can be found in flat or footed bowls, scarce plate shape and the rare rose bowl. The edges often are varied in ruffling and can resemble a jack-in-the-pulpit shape. Colors are marigold, blue, green, white, vaseline, aqua and red.

HORSESHOE SHOT GLASS

If it weren't for the pretty horseshoe design in the bottom of this shot glass, it would be just another piece of glass. But with the design, it is very collectible, highly sought and not cheap. Marigold is the only color so far reported.

HOT SPRINGS VASE

This very interesting vase is a beauty. It measures 9⅞" tall and 4½" wide across the top. The coloring is pale amber and the iridescence is quite good.

It is lettered:

> "To Lena from Uncle
> Hot Spring, Arkansas
> Superior Bath House
> May 20, 1903"

Who says Carnival glass was first made in 1907?

IDYLL (FENTON)

While I apologize for the poor quality of the picture, I have only praise for this very rare vase. It is one of only two ever reported and is a design jewel. The coloring is amethyst, and it stands 7" tall. The pattern of water lillies and butterflies is just right.

ILLINOIS DAISY

How strange the name of this pattern sounds on a British pattern, but Illinois Daisy was a product of Davison of Gateshead. Shapes known are the familiar covered jar and an 8" bowl. The only color I've heard about is marigold, often very weak, but perhaps time will turn up another color.

ILLUSION

It is rather difficult to describe this seldom-discussed Fenton pattern. It is found on bowls occasionally, but mainly on bonbons and is a combination of flowers, leaves and odd geometric shadow-like blotches. Colors seen are marigold and blue, but I certainly would rule out green amethyst. A red Illusion bonbon would be a real treasure, but so far I haven't heard of one.

Idyll (Fenton)

108

Honeycomb and Hobstar

Honeycomb Rose Bowl

Horseshoe Shot Glass

Horses' Heads

Horn of Plenty

Hot Springs Vase

Illinois Daisy

Illusion

IMPERIAL BASKET

This Imperial handled basket is identical in size and shape to the Daisy basket and as you can see, the handl patterns are the same. The coloring is unusual in that it is mostly smoke Carnival with just a touch of marigol around the top.

IMPERIAL JEWELS CANDLE HOLDERS

These beautiful Imperial Jewels Candle Holders are 7¼" tall. As you can see, they have a bold red iridize finish, and there is no stretch effect. Some are marked with the "Iron Cross" mark, but these are not.

IMPERIAL PAPERWEIGHT

This very rare advertising paperweight was shown in my book on Carnival glass rarities for very few of thes are known. The paperweight is roughly rectangular, 5½" long, 3⅛" wide and 1" thick. A depressed oval is in th center of the weight and it is inscribed: "Imperial Glass Company, Bellaire, Ohio, USA," "Imperial Art Glass, "Nucut," "NuArt," and "IM | PE." These are all trademarks of the company, of course. The glass is a fine amethys
   RI | AL
with good, rich luster.

INCA VASE

Like its sister design, the Sea Gulls vase, this is a foreign pattern also. It's probably Scandinavian. The feelin is Art Deco and very well done. The only color reported is a rich marigold.

INDIANA STATEHOUSE

Besides the two versions of the Soldiers and Sailors plates, Fenton also made this very rare plate shown i blue. The size is identical to the other two and the exterior carries the same Berry and Leaf Circle design. Thi was made in marigold also.

INTAGLIO DAISY

Again, the English have designed a neat, useful near-cut pattern. The bowl stands 4" tall and has a 7½" width The design is all intaglio, and the only color I've seen is marigold. Sowerby is the maker.

Imperial Basket

Imperial Jewels Candle Holders

Imperial Paperweight

Inca Vase

Indiana Statehouse

Intaglio Daisy

INTERIOR FLUTE

Here's another one of the many flute patterns so popular in their day. There was probably a matching suga and I haven't seen any other colors but they probably exist.

INTERIOR RAYS

The three-piece table set, consisting of covered butter dish, covered sugar and covered jam jar came fro. Australia. Whether it was made there, I can't say, but I'd guess England as its original home.

INTERIOR RIB (IMPERIAL)

This very classy 7½" vase is tissue paper thin, and the only design is the inside ribbing that runs the leng of the glass. The color is a strong smoke with high luster, and I've seen only two of these over the years.

INTERIOR SWIRL AND WIDE PANEL

Despite its long name, this squat pitcher is a real cute one with good coloring and lots of character. Oth colors and shapes may exist, but I haven't seen them, and I don't know who the maker is.

INVERTED COIN DOT

The tumbler shown is part of a scarce water set made by the Fenton company, and to the best of m knowledge is not found in other shapes. As you can see, the pattern is all interior. Colors known are marigol amethyst and green, but blue may exist.

INVERTED FEATHER

This is probably the best known of all Cambridge Carnival glass patterns and is found in a variety of shape including a cracker jar, table set, water set, compote, sherbet, wine, milk pitcher and punch set. All shapes ar rare except the cracker jar, and colors known are marigold, green and amethyst.

INVERTED STRAWBERRY

Like its cousin, Inverted Thistle, this beautiful Cambridge pattern is all intaglio and can be found on severa shapes including berry sets, water sets, candlesticks, large compotes, sherbets, milk pitchers, creamers, spooner a stemmed celery, powder jars and a lady's spittoon. All shapes are rare and colors known are marigold, gree amethyst and blue.

INVERTED THISTLE

Cambridge was responsible for very original patterns and superior workmanship and here is a prime exampl Known in water sets, a spittoon, a covered box, a pickle dish and a breakfast set, this is an intaglio patter Colors are marigold, amethyst, green and a rare blue.

IRIS

Along with the plain Buttermilk Goblet, this pattern found in both compotes and goblets, is Fenton. Color seen are marigold, green, amethyst and a very rare white.

IRIS AND HERRINGBONE

Made late in Carnival history, this Jeanette pattern can be found in a host of shapes, all useful. Its colorin runs from good to awful, and many pieces are found in crystal also.

Iris and Herringbone

112

Interior Swirl and Wide Panel

Interior Rays

Inverted Coin Dot

Interior Flute

Inverted Feather

Inverted Strawberry

nverted Thistle (Late)

Interior Rib

Iris

ISAAC BENESCH BOWL

This cute 6¼" advertising bowl is quite easily identified because of its distinct design. It bears the labeling "The Great House of Isaac Benesch and Sons, Wilksbarre, Pa., Baltimore, Md., Annapolis, MD." The center theme is bracketed by springs of daisy-like blossoms and leaves. The exterior carries the familiar wide panel design and a rayed base. The predominant color is amethyst.

IVY

This cute little souvenir wine has the same stem and base design as Fenton's small Orange Tree compote and we are sure it came from that factory. It is found in both the wine and claret size with various advertisement. Marigold is the only color I've seen but certainly others may have been made.

JACK-IN-THE-PULPIT

While both Dugan and Fenton made Jack-in-the-Pulpit vases, the one shown is a Northwood pattern and so marked. It stands 8¼" tall and has an exterior ribbing. Colors known are marigold, purple, green, aqua opalescent and white but others may exist.

JACK-IN-THE-PULPIT (DUGAN)

Besides the Northwood version of this vase, Dugan had a try at it too and here is their version. It's a bit plainer and isn't nearly as pretty without the footing but it does come in several colors and adds a pleasant touch of color and artistry.

JACKMAN WHISKEY

Much like its cousin, the Golden Wedding bottle, this was designed to hold whiskey for commercial sale. It scarcer than the former and generally has better iridescence. The maker is unknown.

JACOB'S LADDER VARIANT (U.S. GLASS)

Besides the Jacob's Ladder perfume, I've heard of this rose bowl shape in the variant and nothing else. Just why more shapes haven't surfaced, I don't know, for the pattern is interesting and well done. The only color is a good marigold.

JELLY JAR

For years I found the lids to these Imperial Jelly Jars in shops and thought they were late Carnival glass "coasters," but a few years back, I saw the two parts put together and was really quite surprised at what I saw. The jar itself is 3" wide and 2¾" tall and of deep, well iridized marigold. All the pattern is interior so that when the jar was upended onto the lid, a design was formed in the jelly. The lid itself also carries an interior design of spokes and an exterior one of a many-rayed star. Again, here is a true rarity, well within any collector's range.

JEWELED HEART

Carried over from the pressed glass era, Jeweled Heart is, of course, the famous exterior pattern of the Farmyard bowl. However, Dugan used it as a primary pattern on very scarce water sets when the pitcher footed – an uncommon shape for water pitchers in Carnival glass. I have seen Jeweled Heart in purple, peach and marigold but other colors may exist. If so, they would be considered ultra-rare. A tumbler is reported in white, but I haven't seen it.

JEWELS

This was the name given to a line of unpatterned iridized glass made by the Imperial company at the same time they were making their wonderful Carnival. Whether you consider it Carnival glass is your business, but I do, so I'm showing it here. It was made in a multitude of shapes and most Imperial colors.

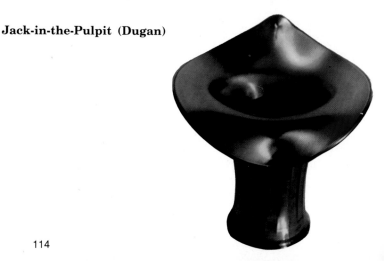

Jack-in-the-Pulpit (Dugan)

Jacob's Ladder Variant (U.S. Glass)

Isaac Benesch Bowl

Jack-in-the-Pulpit

Jackman Whiskey

Ivy (Wine)

Jelly Jar

Jeweled Heart

Jewels

JOCKEY CLUB

Although very much akin to the Good Luck pattern, Northwood's Jockey Club is certainly a separate patter and a quick once-over will establish this fact. The floral arrangements are entirely different, even the horsesh and riding crop are not the same. Jockey Club is found on trademarked bowls which carry the Northwoc Basketweave as an exterior pattern. The ones I'm familiar with have been on a good amethyst glass, well iridize

KANGAROO

This beautiful Australian pattern typifies the quality of the Crystal Glass Factory. Known in large and sma bowls, in marigold and purple, Kangaroo has the Wild Fern pattern as its exterior. There is a variant.

KEYHOLE (DUGAN)

Used as the exterior pattern on Raindrop bowls, this Dugan secondary pattern is much more interesting tha the one it compliments. Colors are marigold, amethyst and peach opalescent.

KEYSTONE COLONIAL (WESTMORELAND)

At least three of these have been reported, all with the keystone trademark containing a "W." They stan 6¼" tall and are very much like the Chippendale pattern compotes credited to Jefferson Glass. The only color I'v seen is a very dark purple.

KINGFISHER

Like many of the other "bird patterns" from down under, this nice Australian bowl has both the animal an flora shown. Kingfisher has a stippled aura, a registration number, and comes in large and small bowls.

KINGFISHER VARIANT

The primary difference between this and the regular Australian Kingfisher pattern is the addition of a wreat of wattle. Both marigold and purple are found.

KING'S CROWN (U.S. GLASS)

In crystal and ruby flashed glass, this pattern can be found in many shapes, but this is the only piece i Carnival glass I've ever heard about. It is a dainty wine goblet in good marigold and very, very rare indeed.

KITTEN PAPERWEIGHT

This slight bit of glass is a rare miniature heretofore unreported in any of the pattern books on Carnival glas Its coloring, as you can see, is a good even marigold, and the iridization is well applied. In size, it is much lik the 3" long Bulldog paperweight. The Kitten paperweight is truly a rare little item.

KITTENS

For many years, this Fenton pattern has been a favorite of collectors and the Kittens pieces rise steadily i price and popularity. This is a child's pattern and is known in bowls, cup and saucer, banana bowl, spooner, plat vase and a very are spittoon whimsey. Colors are marigold, vaseline and cobalt blue.

King's Crown (U.S. Glass)

Keystone Colonial (Westmoreland)

Jockey Club

Kangaroo

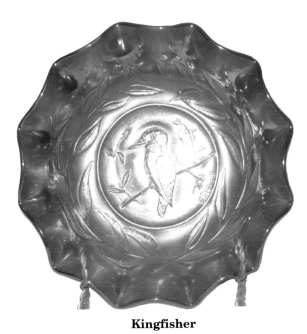

Kingfisher

Kingfisher Variant

Kitten Paperweight

Kittens

Keyhole (Dugan)

KIWI

This very odd bird is a native of New Zealand. This Australian bowl shows two of them, with a mountain rang
in the background and a border of fine fern branches. Colors are a fine purple and marigold.

KNOTTED BEADS

In many ways, this is a combination of the April Showers vase pattern and the Diamond and Rib vase patter
for the tiny grouping are gathered into pulled ovals. I know of examples of Knotted Beads in marigold, blu
green, white and vaseline, and sizes range from 4" to 12" in height. A Fenton product.

KOOKABURRA

Similar to the Kingfisher pattern, the Australian Kookaburra has an upper edging of Flannel Flower, wattl
and two large groups of Waratah flowers flanking a Flannel Flower blossom.

KOOKABURA VARIANT

In this Australian variant, the obvious differences are the fine large Waratah flowers around the stipple
circle. And, of course, the edging is a bullet rim pattern.

LBJ HAT

This late Carnival ashtray has been called this for the last few years. It has a diameter of 6¼" and stand
3" tall. While not in the class of a Cleveland Memorial Ashtray, it is an attractive specimen of iridized glass

LACY DEWDROP

Lacy Dewdrop appears in iridized glass in one way only – the pearl Carnival shown. Shapes reported ar
pitchers, tumblers, covered compotes, covered bowls and banana bowls. Westmoreland is the possible maker.

LATE WATERLILY

For one of the later patterns in Carnival glass, this certainly is a good one. The coloring is superior, the shap
symmetrical, and the design of cattails and waterlilies attractive.

Late Waterlily

Kiwi

Knotted Beads

Kookaburra

Kookaburra Variant

LBJ Hat

Lacy Dewdrop

LATTICE AND DAISY

Most often seen on marigold water sets, Dugan's Lattice and Daisy was also made on very scarce berry set The other colors known are cobalt blue and white, but I certainly would not rule out green or amethyst, especial on the water sets.

LATTICE AND GRAPE

If you examine the shape of this Fenton water set, you'll find it is almost identical to that of the Grapevir Lattice water set, but Lattice and Grape is distinctive in its own right. The mold work is very good and the desig appealing. Available in a rare spittooon whimsey pulled from a tumbler in addition to water sets, Lattice an Grape is found in marigold, cobalt blue, green, amethyst, peach opalescent and white.

LAUREL AND GRAPE

I know very little about this previously unreported pattern except that it is a vase shape and stands 7" ta The design, a cluster of grapes inside a laurel wreath is a nice one and reminds me of the English Grape an Cherry pattern. Marigold is the only reported color.

LEA

Found in a handled pickle dish, a footed creamer, and a footed bowl that can be in various shapes, Lea an the Lea variants are patterns produced by the Sowerby company. Most examples are marigold, but I have hear of an amethyst creamer.

LEAF AND BEADS

Here is another very well-known Northwood pattern found not only in Carnival glass, but clear and opalescer glass as well. Available on bowls as well as rose bowls, the twig feet are used by more than one Carnival glas maker. However, the trademark is usually present. Leaf and Beads is found in a wide range of color includin marigold, green, purple, ice green, ice blue and white.

LEAF AND BEADS VARIANT

Here is the variant bowl on a dome base and as you can see, the design is not so strong as the regular Lea and Beads. These seem to be found mostly in green, as shown, but marigold and amethyst are known. It i marked Northwood. The shape is considered a nut bowl by many collectors.

LEAF AND LITTLE FLOWERS

This little Millersburg cutie is rather unique, not only in size but also in design. The four free-floating bloom are similar to those on the Little Stars design, but the four large prickly-edged leaves with a center cross blossor is unique. It seems cactus-like. Just 3" tall, the Leaf and Little Flowers compote, has an octagon-shaped base an has been seen in amethyst, green and rarely marigold.

LEAF CHAIN

Very close to Cherry Chain in design, Fenton's Leaf Chain is more commonly found, indicating it was mad in larger amounts. Known in bowls, plates and bonbons. Leaf Chain was made in a wide variety of color including marigold, cobalt blue, green, amethyst, white, vaseline, lavender, smoke, aqua and red. The mold wor is quite good and the finish above average. Shown in a rare marigold opalescent.

Leaf and Beads Variant

Laurel and Grape

120

Lattice and Daisy

Lattice and Grape

Lea

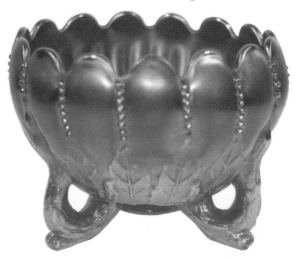

Leaf and Beads

Leaf and Little Flowers

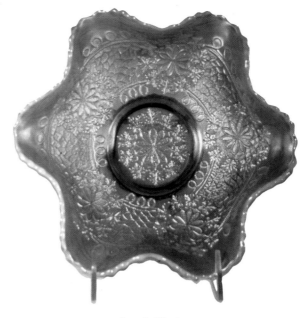

Leaf Chain

LEAF COLUMN

While not a spectacular pattern, here is a Northwood vase that takes on new importance on second glance. It is a well-balanced all-over pattern that does the job nicely. The iridescence, especially on the dark colors, is quite nice. The shape is attractive and the vase is, of course, a very useful item. Certainly any collection of Carnvial glass would benefit by adding one of these.

LEAF RAYS

Found mostly on a one-handled, spade-shaped nappy, this Dugan pattern was also made in the ruffled version shown. Colors are marigold, peach opalescent, amethyst, green, blue, white and clear.

LEAF SWIRL

Here is a much overlooked pattern and that certainly is a shame because it is a beautiful, well-executed one. This compote is almost 5" tall and has three mold marks. It is usually found on a very deep shape of purple but we are delighted to show you this teal color so very rare. Amber and marigold are also known. A Westmoreland product.

LEAF SWIRL AND FLOWER

What a beautifully graceful vase this is. It appears to be a Fenton product, stands 8" tall and has a trail of etched leaves and flowers around the body. I'd guess it was made in marigold and pastels too, but can't be sure.

LEAF TIERS

For some reason, this pattern is difficult to locate, so apparently it wasn't made in large quantities although it was made in useable shapes, including berry sets, table sets and water sets. The twig feet were used by both Fenton and Northwood to some degree, and while Leaf Tiers is mostly seen in marigold, very scarce water sets in purple, green and blue are known. A Fenton product.

LILY OF THE VALLEY

Personally, I feel this is the very best of all Fenton water sets, and the price this rare pattern commands bears me out. Aside from being beautiful, Lily of the Valley is an imaginative pattern. Found only in cobalt blue and marigold, Lily of the Valley would have made a beautiful table set pattern.

Leaf Column

Leaf Rays

Leaf Swirl

Leaf Swirl and Flower

Leaf Tiers

Lily of the Valley

LINED LATTICE

Most often seen in a vase shape with odd feet-like projections, the same design is used on a shade frequentl associated with the Princess lamp base. The vase, we know, was made by Dugan and can be found in marigold amethyst, blue, green, peach opalescent and white.

LION

Someone at the Fenton factory had to be an animal lover for more animal patterns were born there than a any other Carnival glass factory in America. The Lion pattern is a nice one, rather scarce and available in bowl and plates. The colors are marigold and blue in the bowls, but I've seen only marigold in the rare plates.

LITTLE BARREL

I have no proof that this is an Imperial design; however, the luster, mold work and the color of helios green marigold, amber and smoke all suggest the Little Barrel is an Imperial product. They are rumored to have hel a sample of some liquid and examples have been found with paper labels on the exterior suggesting a giveawa item. Nonetheless, they are hard to find, very collectible, and much loved.

LITTLE BEADS

Mady by the Westmoreland company, this odd little pattern is seen mostly on peach opal, but the exampl shown has an aqua base glass. It measures 2" tall, 5½" across and has a 2½" base.

LITTLE DAISIES

Apparently Fenton made very small amounts of this pattern for few examples are to be found. Margold is th only color reported and the only bowl shape found. The exterior is plain.

LITTLE FISHES

A close comparison with the Coral pattern will reveal many similarities to this pattern. Little Fishes can b found in large or small bowls, both flat based and footed, and in rare plates that measure about 9". Colors ar marigold, blue, green, amethyst, aqua, vaseline, amber, ice green and white, which is rather rare.

Lined Lattice

Lion

Little Barrel

Little Beads

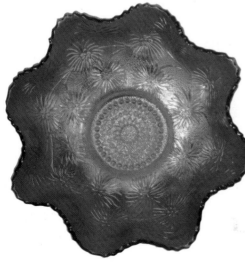

Little Daisies

Little Fishes

LITTLE FLOWERS

Here's another pattern once felt to be Millersburg but now known to be a Fenton product. Little Flowers found in berry sets and two sizes of rare plates, in marigold, green, blue amethyst, amber, aqua, vaseline and red.

LITTLE STARS

I've found this Millersburg pattern a bit difficult to locate and don't believe many of these bowls are around. The amethyst ones are especially nice, but green and marigold exist. Also a larger size is known as is the plate shape in green. Shown is a rare blue bowl.

LOGANBERRY

Despite the reproduction of this impressive vase in the 1960's, Imperial's Loganberry has remained one of the collector's favorites. The vase is 10" tall and has a base diameter of 3¾". There are four mold marks. The colors found are green, marigold, amber, smoke and purple with the latter hardest to find. Of course, the quality of design and the beautiful luster make this a real treasure.

LONG HOBSTAR

Imperial was famous for near-cut designs, and Long Hobstar is a prime example of their skills. Shown in very rare punch bowl and base (no cups known), but Long Hobstar can also be found in bowls, and a beautiful Bride's Basket. The colors are marigold, smoke, green and occasionally purple.

LONG THUMBPRINT

Found in bowls, compotes, creamers, sugars and vases, this Dugan pattern has little going for it design-wise but sometimes is quite nice when the coloring is as good as on the vase shown. Colors are mostly marigold, but amethyst, blue and green are occasionally found and I've seen the breakfast pieces in a smoky marigold, and the vase in peach opalescent.

LOTUS AND GRAPE

Please note the similarities between this pattern and Fenton's Water Lily and Two Flowers. All have a feeling of the small ponds once so much a part of backyard garden displays. Lotus and Grape is found on bonbons, flat or footed bowls, and rare plates in colors of marigold, green, blue, amethyst, white and red.

LOTUS LAND

What a privilege to show this very rare bonbon, generous in size (8¼" across) and rich in design. From its stippled center flower to the whimsical outer flowers, the pattern is one you won't soon forget. Amethyst is the only reported color on the few examples.

Lotus Land

Little Flowers

Little Stars

Loganberry

Long Hobstar

Long Thumbprint

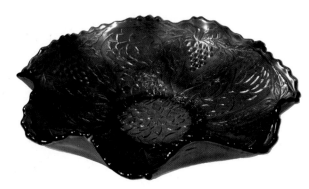

Lotus and Grape

LOUISA

Louisa is a very pretty, well designed floral pattern that seems perfectly suited for the rose bowl shape show. Also known on footed candy dishes and a rare footed plate (all from the same mold), Louisa is found in marigol green, blue, amethyst, and the rare deep amber called "horehound." This is by Westmoreland.

LOVELY

Northwood's Lovely is a seldom-seen interior pattern found on footed bowls with Leaf and Beads as an exteric pattern. While we can't be certain, it is possible this motif was added at the Dugan factory since shards of Lea and Beads were found there. The colors seen thus far are marigold and purple with very good iridescence.

LUSTRE AND CLEAR (PILLAR AND FLUTE)

Most often seen in footed creamer and open sugar, I've seen this pattern on a beautiful Clambroth rose bov (marked with the old "Ironcross" mark), bowls of several sizes, a celery tray, water sets, a compote and a fan vas. Originally called Pillar and Flute (#682), this Imperial pattern was made in many, many shapes in crystal. Th usual colors are marigold or smoke with an occasional green or purple piece appearing.

LUSTRE FLUTE

Again we have a very familiar pattern to most collectors, but one that is not really too distinguished. I suppos not every pattern should be expected to be spectacular. The shape I've seen most often is the hat shape in bot green and marigold, but punch sets, berry sets, breakfast sets, bonbons, nappies and compotes do exist i marigold, green and purple. The base is many-rayed and usually the Northwood trademark is present.

LUSTRE ROSE

Aside from Imperial's Grape or Pansy patterns, this is probably their biggest seller in the early years and wa made in berry sets, table sets, water sets, plates, fruit bowls, and a beautiful fernery in colors of marigold, gree purple, cobalt, amber, smoke and clambroth. Reproduction has devalued this beautiful pattern, but the old piece are still outstanding examples of glass making, with superb craftsmanship and color.

MADAY AND CO.

This quite rare advertising bowl is found as the exterior of Fenton's Wild Blackberry 8½" bowl. The only colc known to date is amethyst, while the normal bowls are found in green and marigold primarily. The advertisin reads "H. Maday and Co. 1910."

MAGPIE

The Magpie is a different bird altogether and the one shown on this Australian bowl is probably a Ne Zealand Parson bird. The flowers are tyical Flannel Flowers and wattle.

Maday and Co.

Louisa

Lovely

Lustre and Clear
(Pillar and Flute)

Lustre Flute

Lustre Rose

Magpie

MALAGA

Malaga is rather difficult pattern to find, indicating production must have been small on this Dugan desig
What a pity for the all-over grape pattern is a good one with imaginative detail throughout. Found only on lar
bowls and plates, I've heard of marigold, amber and purple only, but green is a strong possibility.

MALLARD DUCK

I certainly wish I knew more about this rare item, but it's the only true Carnival one I've seen. I believe the
were made by Tiffin (mostly in milk glass), and I once saw an example with applied ruby luster. However, I ca
say with assurance the one shown is old, has been in one of the country's major collections for years and is
prized rarity of the owner. The coloring is a beautiful clambroth with fiery blue and pink highlights.

MANY FRUITS

This is a truly lovely fruit pattern, something that any company would be proud to claim. The mold work
heavy and distinct, the design is interesting and quite realistic and the coloration flawless. I personally pref
the ruffled base, but that is a small matter. This Dugan pattern would have made a beautiful water set. Th
colors are marigold, blue, white, purple and green.

MANY STARS

Which came first – the chicken or the egg? Or in this case, the Many Stars or the Bernheimer bowl? For th
are exactly the same except for the center design where in the former a large star replaces the advertising. The
bowls are generous in size and can be found in amethyst, green, marigold and blue. The green is often a ligh
airy shade just a bit darker than an ice green and is very attractive when found with a gold iridescenc
Millersburg manufactured this pattern.

MAPLE LEAF

Maple Leaf is a carry-over pattern from the custard glass line, but in Carnival glass is limited to stemme
berry sets, table sets and water sets. I examined shards of this pattern from the Dugan dump site, so items
Carnival glass were obviously turned out at that factory. The background is the same Soda Gold pattern as th
found on the exterior of Garden Path bowls and plates. Maple Leaf was made in marigold, purple, cobalt blu
and green.

MARILYN

This water set pattern is probably one of the most unusual and outstanding in the field of Carnvial glass. Fir
look at the pitcher's shape. Notice the unusual upper edging, so different than those of other companies. The
there is the drooping pouring lip so favored by the Millersburg company. The finish is, of course, the fine radiu
look and the glass is heavy. All in all, a real prize for any collector.

MARY ANN

I don't believe I've heard of another one of these loving cups in collections I'm familiar with, and yet I'
always known these existed. As you can see, they are much different from the vase shape, having three handl
and no scalloping around the top edge or base. Of course, it is quite rare, and it is a real pleasure to show
here. The Mary Ann vase is Dugan.

MASSACHUSETTS (U.S. GLASS)

While the vase shape shown is the only shape I've heard about in this very attractive pattern, I'd bet it wa
made in a creamer and sugar as well. It is found only in marigold and brings a good price when sold for it
a scarce and desirable item.

Mary Ann

130

Malaga

Mallard Duck

Many Fruits

Many Stars

Maple Leaf

Massachusetts

Marilyn

MAY BASKET

This beautiful basket is called Diamond and Fleur de lis by Mrs. Presznick. It was made by Davisons Gateshead and has a diameter of 7½" and is 6" tall. Can you imagine how difficult it was to remove this beautif novelty form the mold? May Basket was made in marigold and has been reported in smoke.

MAYAN

Mayan is a quite unique pattern; once seen, it is easily identified. Its six feather-like plumes stand out proud from the raised, beaded button center. Of course, the beaded border is a Millersburg favorite that appears on th Daisy Square pedestal rose bowl as well as the Olympic compote.

MAYFLOWER (MILLERSBURG)

Found with the rare Millersburg Grape Leaves bowl, Mayflower is a series of flower-like designs, separate by eight diamond and near-cut sections. Held to the light, the effect is a real surprise through the Grape desig Colors are marigold, green, amethyst and vaseline and every one is rare.

MEANDER (NORTHWOOD)

Here's the exterior of the gorgeous Three Fruits Medallion bowl shown elsewhere, and as you can see, the gla is truly black amethyst and has no iridizing on the exterior. The pattern of Meander, however, is a good desig and very complimentary.

MELON RIB

Another pattern credited to Imperial, Melon Rib can be found on water sets, candy jars, salt and peppe shakers and a covered powder jar, all in marigold. As you can see, the pitcher is tall and stately and very prett

MEMPHIS

While an interesting geometrical pattern, Northwood's Memphis has never been one of my favorites – possib because of its limited shapes. I can imagine how much my interest would increase if a water set were to appea I have seen an enormous banquet punch set in crystal, but the size was not made in Carnival glass. The shape are a berry set, punch set, fruit bowl on separate stand, and a compote. The colors found are both vivid an pastel.

MIKADO

Called giant compotes, these beautiful 8½" tall fruit stands have two outstanding patterns – cherries on th exterior and the beautiful Mikado on the inside of the bowl. This design consists of a center Medallion of stipple rays around which are three large chrysanthemums and three oriental scroll devices. The colors are marigol blue, a rare green and a very rare red. Mikado was made by Fenton.

MILADY

Again, we show one of the better water sets by the Fenton company. The pitcher is a tankard size, and th design is paneled with very artistic blossoms and stems with graceful leaves. Colors most seen are marigold an blue, but scarce green and a rare amethyst do exist as shown.

Mayflower (Millersburg)

Meander (Northwood)

Melon Rib

May Basket

Mayan

Memphis

Mikado

Milady

MINIATURE BLACKBERRY COMPOTE

This little Fenton cutie is most often seen on blue or marigold and is quite scarce in either color. The whit one shown is very rare. Standing only 2½" high with a bowl diameter of 4⅛", this miniature or jelly compote woul be a treasure in any collection and certainly would highlight a compote grouping.

MINIATURE HOBNAIL CORDIAL SET

What a nice little set this is! Probably from Europe, it is the only one I've heard about. The decanter stand 8" tall to the top of its stopper and needless to say, the set is quite a rare item.

MIRRORED LOTUS

Besides a rare white rose bowl, Fenton's Mirrored Lotus is most seen in a 7" ruffled bowl in blue, gree marigold or white. A bonbon shape and a plate shape are also known.

MITERED DIAMONDS AND PLEATS

This British pattern is most often seen as shown, a 4½" handled sauce dish, but I've seen 8½" bowls and Mr Presznick reports a tray in 10" size as well. The coloring is top notch marigold, but a smoky blue shade has bee reported.

MITERED OVALS

This beautiful vase is a Millersburg product. It is a rare item seldom found for sale. The colors seen ar amethyst, green and very rarely marigold. In size, it is outstanding, being some 10½" tall. The mold work i superior.

MODERNE

I've named this cute cup and saucer the very first name that entered by head, but it seems to fit. I'd gues it is fairly late Carnival, but it still has good coloring and a nice luster.

MOONPRINT

Here is one of the prettiest of all English Carnival patterns. The shapes are a banana bowl, covered jar, vas covered butter, covered cheesekeeper, compote, milk pitcher, creamer and bowls that range in size from 8" to 15 Marigold is the basic color wtih a super finish. I've heard of a peach opalescent covered jar but cannot confirm its existence. I'd guess Sowerby made this pattern.

MORNING GLORY

If there is one water set that stands above all others in sheer beauty, this Millersburg pattern is surely i Almost 14" tall, the pitcher is stately. The heavily raised morning glory vines around its center and the applie handle has been shaped to resemble a leaf where it joins the pitcher. The matching tumbler is 3¾" tall and ha a rayed base. The glass is sparkling clear, the radium finish is excellent and the mold work impressive. Alon with the Gay 90's, the Morning Glory water set has to be near the top of anyone's list.

Moderne

Miniature Hobnail Cordial Set

Miniature Blackberry Compote

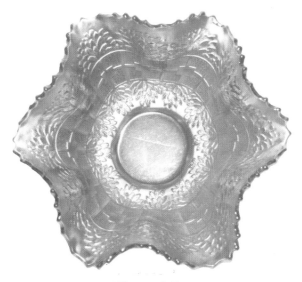

Mirrored Lotus

Mitered Diamonds and Pleats

Mitered Ovals

Moonprint

Morning Glory

MORNING GLORY VASE

Several companies made similar vases with the flaring top resembling morning glories, but the one show happens to come from Imperial and is a whopping 17" tall. The color is fabulous! I've seen these as small as 8 and every size between, in several colors including marigold, smoke, green, blue amethyst, white, ice green, tea and aqua . . . even one in a strange pale blue and marigold combination.

MT. GAMBIER MUG

I can't be sure this mug is Australian, but this one comes from there. It is etched "Greetings from M Gambier." The color is a good rich marigold.

MOXIE BOTTLE

I normally avoid the bottle cycle unless they are attractive as this well-designed scarcity. The coloring obviously a very frosty white with heavy luster, and the design of plain and stippled diamonds is a good on Moxie was a soft drink that went out of favor in the mid 1920's.

MULTI-FRUITS AND FLOWERS

Grape clusters, leaves, blossoms, cherries, peaches and pears! What an imaginative collection to grace beautif lustered glass! This fantastic Millersburg pattern is seen mostly on punch sets of medium size, but here is a ver scarce water set and a stemmed fruit goblet or compote. The base of the punch set can double a a compote whe up-ended and is iridized inside and out.

MY LADY'S POWDER BOX

Here is a favorite with collectors and a good price is always assured when one of these sell. Found only i marigold of a good rich quality, the powder jar stands 5½" tall and has a base diameter of 3¼". The figure o the lid is of solid glass. I suspect Davisons is the manufacturer, but have no proof.

MYSTIC

I named this 7" vase so if anyone knows it by another name, I'd be happy to hear from them. The design elongated shields separated by a hobstar and file filler is interesting. The very mellow coloring reminds me c other Cambridge products, and that is just who made it, according to a 1908 advertisement.

NAUTILUS

It is a shame there aren't more shapes of this pattern in Carnival glass because every piece of the custar shapes are a joy. In Carnival we find only the small novelty piece described as a boat shape (actually one larg footed berry bowl has been seen in marigold – a very, very rare piece of Carnival glass! It appears with eithe both ends turned up or one end turned down, and the colors are peach and purple. A Dugan product from ol Northwood molds.

NEAR-CUT

This attractive Northwood pattern is quite similar to the Hobstar Flower shown elsewhere in this book. In addition to the compote and goblet, made from the same mold, there is a very rare water pitcher. I have heard of no tumblers but they may exist. The colors are marigold and purple with the latter most seen. A rare green compote is known also.

Mystic

Mt. Gambier Mug

Moxie Bottle

Multi-Fruits and Flowers

My Lady's Powder Box

Nautilus

Morning Glory

Near-Cut

137

NEAR-CUT DECANTER

What a triumph of near-cut design this rare Cambridge pattern is! Standing 11" tall, this beauty is found in a sparkling green, but I wouldn't rule out amethyst or marigold as possibilities. The mold work is some of the best, and the finish is equal to any from the Millersburg plant.

NESTING SWAN

This pattern has become one of the most sought Millersburg bowl patterns, and it is certainly easy to see why. The graceful swan, nesting on a bed of reeds, surrounded by leaves and blossoms and cattails is a very interesting design. The detail is quite good and the color and finish exceptional. The beautiful Diamond and Fan exterior contrasts nicely and echoes the fine workmanship throughout. In addition to the beautiful green, marigold and amethyst, Nesting Swan can be found in a beautiful honey-amber shade and a very rare blue.

NEW ORLEANS CHAMPAGNE

Like the other champagnes, from U.S. Glass, this version is on clear iridized glass with the design areas hand painted. The alligators along the sides are quite unique in Carnival glass and the crowned and bearded man represents Rex, King of the Mardi Gras. I really can't think of anything more fitting for a piece of Carnival glass.

NIGHT STARS

What a rare little beauty this Millersburg pattern is! Found on the bonbon shape shown in an unusual olive green, marigold and amethyst, it has also been seen in a very rare card tray in amethyst and vaseline and an equally rare spade-shaped nappy in amethyst only. All shapes and colors are rare and desirable so never pass one up.

NIPPON

Certainly Nippon must have been a popular pattern in its heyday, for it is readily found today. The pattern is simple but effective; a central stylized blossom with panels of drapery extending toward the outer edges of the bowl. Found in a wide variety of colors, including marigold, green, blue, purple, ice green, ice blue and white. This Northwood is a nice pattern to own.

NORTHERN STAR

I have some misgivings about this being a Fenton pattern, but since it is listed as such by a noted glass author, I will consider it one. Shown is the 6" card tray, but Northern Star can also be found on a ruffled mint dish, small bowls and plates, all in marigold. The design is all on the exterior and both outside and inside are iridized.

Near Cut Decanter

Nesting Swan

New Orleans Champagne

Night Stars

Nippon

Northern Star

NU-ART CHRYSANTHEMUM

Apparently the popularity of the regular Nu-Art plate prompted this, a sister design of striking beauty. Al 10½" in diameter, the Chrysanthemum plate has the same Greek key border device. The flowers are very gracef and heavily raised and, of course, the iridescence is outstanding. I've seen this plate in marigold, smoke, ambe green, clambroth and purple, but other colors may exist. Again, this pattern was reproduced in the 1960's. Th was produced by Imperial.

NU-ART (HOMESTEAD)

If you will take a few moments to compare this Imperial pattern with the Double Dutch found elsewhere this book, you will find very similar designs, probably done by the same artist. The Nu-Art plate, however, is rarer more important design that sells for many times the amount of a Double Dutch bowl. Found is many colo including marigold, green, purple, smoke, amber, white, ice green and the rare cobalt blue, the Nu-Art plate sometimes signed. It measures 10½" across and has been reproduced.

NUGGATE

This very attractive handled bottle is 4½" tall and 4" at the widest diameter. The coloring is a super coba but the ribbed handle is marigold! I don't know much about it except it is European and very pretty.

#5

This pattern dates from the early days of the Imperial company and was first issued in crystal. The shape known are a celery holder, 6" tall, and the beautiful dome-footed bowl shown, where the pattern is exterior. Th color most encountered is marigold, but as you can see, the bowl is a rich amber.

#4

Made first in crystal, like so many Imperial patterns, this is a rather simple, not too impressive pattern, four on compotes as well as small footed bowls. The colors are usually marigold or smoke but I've seen it in cle Carnival as well as green.

#9

This little Imperial cutie is very nice, especially when found on a rich smoky color with golden highlights. (course, it is also known on marigold, and I suspect green is a possibility. While the pattern is relatively simp – a series of arches filled with small hexogonal buttons – it is quite effective. Sometimes called Tulip and Can #9 is also found in wine, claret, and goblet shapes.

#270

Made by the Westmoreland company, this little open-stemmed bowl is found in peach opalescent milk gla as well as the aqua color shown. It is a simple but effective pattern well worth collecting.

Nuggate

Nu-Art (Homestead)

Nu-Art Chrysanthemum

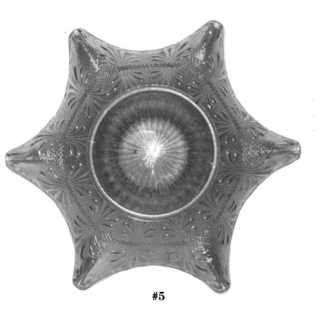

#5

#4

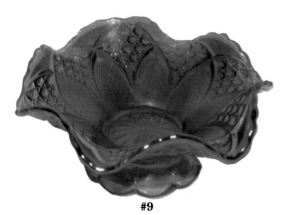

#9

#270

OCTAGON

Next to the Fashion pattern, this is probably Imperial's most common near-cut design, especially when fou... in marigold. But dark colors show up now and then, and the shape shown is rare in marigold. In the beautif... purple, the toothpick is extremely rare. Octagon is found in table sets, water sets, wine sets, footed vases, mi... pitchers, goblets and rare toothpicks. It is a pleasing all-over pattern.

OCTET

Even if this pattern were not marked, we would attribute it to the Northwood company because the exteri... pattern is the Northwood vintage found on the Star of David and Bows bowl. Octet is also a dome-footed bow... usually about 8½" in diameter. It is a simple but effective pattern – one that wouldn't be easily confused wi... others. The colors are marigold, purple, green, white and ice green. The purple is the most common.

OHIO STAR

This beautiful near-cut vase is almost 10" tall and certainly is a standout in the Carnival vase field. Whi... the majority of vases are of simple design, this one flaunts its multi-cut pattern even to the star in the hi... domed base. The coloring is excellent and not only is Ohio Star found in the usual marigold, green, and purp... but is reported in a beautiful blue! Certainly Millersburg blues are not easily found and one of these wou... enchance any collection. A rare compote is known, and a super-rare white vase also.

OLYMPIC COMPOTE

The Millersburg Olympic miniature compote is **extremely** rare. Its measurements are the same as the Le... and Little Flowers compote made by the same company, and the exterior and base are identical also. If ever t... old adage "Great things come in small packages" could apply, certainly it would be to the Olympic compote.

OMNIBUS

This interesting tumbler has been credited to the U.S. Glass company and is rather rare (six known) and litt... recognized. The primary design is a sunburst teamed with two diamonds of file and a pulled diamond of bubbl... like filler, along with fanning. The sunburst moves from top to bottom in a series around the tumbler. The desig... is quite good, the color super. A rare find.

OPEN EDGE BASKET

This rather common Fenton pattern is found quite frequently in several sizes and colors, especially blu... marigold, green and amethyst. But pink, ice blue, ice green, white and red examples do exist in variously shape... items including hat shapes, vase whimseys, banana boat whimseys and bowl shapes. The interior may be pla... or carry the Blackberry pattern and sometimes advertising is present.

Omnibus

Octagon

Octet

Ohio Star

Olympic Compote

Open Edge Basket

OPEN ROSE

This pattern is very similar to the Lustre Rose pattern but is not found on the wide range of shapes as the latter. The plate shown is the most sought shape but there are also footed and flat bowls of many sizes available. Colors of marigold, smoke, green, purple, clambroth and amber are known and each is usually outstanding. The amber plate shown typifies the Imperial quality.

OPTIC VARIANT

If you examine the Optic Flute that follows, you will recognize the base of this Imperial bowl but will notice the fluting is missing. These have been found on berry sets but I'm sure other shapes were made. The interior of the 6" bowl shown was highly iridized and had a stretch appearance and the exterior had only slight lustre.

OPTIC (IMPERIAL)

Each panel in this pattern is curved, thus the name and an interesting design above the ordinary. Besides bowls and small compotes, a creamer and sugar are known in marigold and smoke.

OPTIC AND BUTTONS

In crystal, this pattern is found in many shapes, including table sets, plates, oil bottles, decanters, shakers and sherbets, but in Carnival glass, Optic and Buttons is limited to berry sets, a goblet, a large handled bowl, a small pitcher, tumblers in two shapes and a rare cup and saucer. Many of the items in Carnival are marked with the Imperial "iron cross" mark, including the milk pitcher and the cup. All shapes I've seen are in marigold only.

OPTIC FLUTE

This Imperial pattern is seldom mentioned but can be found on berry sets as well as compotes. Colors I've seen are marigold and smoke but others may have been made.

ORANGE PEEL

This is a sister design to the Fruit Salad pattern and both are made by Westmoreland. Orange Peel, which not as rare, is made in a punch set, custard set and a stemmed dessert in marigold, amethyst and teal, all Westmoreland prime colors.

ORANGE TREE

No other Fenton pattern had more popularity or was made in more shapes than the Orange Tree and all its variants. Known in berry sets, table sets, water sets, ice cream sets, breakfast sets, compotes, mugs, plates, powder jars, hatpin holders, rose bowls, a loving cup, wines, punch sets and goblets. Orange Tree is found in marigold, blue, green, amethyst, peach opalescent, lustered milk glass, aqua opalescent, white, amber, vaseline, aqua, red and amberina.

ORANGE TREE AND SCROLL

What a beauty this hard-to-find tankard set is. The Orange Trees are like those on the regular pieces and the Orange Tree Orchard set, but below the trees are panels of scroll work much like the design on the Milady pattern. Colors are marigold, blue and green but I wouldn't rule out amethyst or white on this Fenton product.

Optic Variant

Optic (Imperial)

144

Open Rose

Optic and Buttons

Optic Flute

Orange Peel

Orange Tree

Orange Tree and Scroll

ORANGE TREE ORCHARD

Obviously a spin-off pattern of the Orange Tree, this rather scarce Fenton water set has a nicley shape bulbous pitcher. The design is a series of Orange Trees separated by fancy scroll work and has been reported marigold, blue, green, amethyst and white.

ORIENTAL POPPY

Here is a very impressive, realistic pattern, especially effective on the chosen shape – a tankard water set. Th mold work on this Northwood product is clear and clean and the glass is quality all the way. Colors are marigol green, purple, white, ice green, ice blue and blue.

OSTRICH CAKE PLATE

This Australian pattern is actually the Emu on a footed cake stand, but has been misnamed. The exterior a beautiful Rib and Cane pattern. Colors are marigold and purple but these are rare.

OVAL AND ROUND

While Imperial's Oval and Round may certainly be thought of as a very ordinary pattern, it does have its ov charm on a bowl as nicely ruffled as the one shown. Simple in execution, the pattern is found on plates of larg size and bowls of various sizes only. The colors are marigold, green or purple and are usually very richly lustere

PALM BEACH

Apparently this U.S. Glass pattern was carried over from the pressed glass days. It is scarce in all shapes b can be found in a variety of useful pieces, including berry sets, table set, a cider set, rose bowls, vase whimsey a plate and a miniature banana bowl whimsey. The color is often rather weak and marigold, purple and whi are known. The example shown is a bowl whimsey with an iridized "goofus" finish.

PANELED DANDELION

Fenton's Paneled Dandelion is another of those spectacular tankard water sets that are so eye-catching. Th panels of serrated leaves and cottony blossoms are very realistic and fill the space allowed nicely. Colors a marigold, blue, green and amethyst.

Orange Tree Orchard

Oriental Poppy

Ostrich Cake Plate

Oval and Round

Palm Beach

Paneled Dandelion

PANELED HOLLY

This Northwood pattern is found in crystal, gilt glass and Carnival glass. However, the range of shapes much less in the latter, limited to the exteriors of bowls, footed bonbons, a rare breakfast set, and an extreme rare water pitcher. While fairly attractive, Paneled Holly is really not a great pattern. It appears a trifle bus a bit confused. The most often seen color is green but purple and marigold do exist, and I've seen a combinatio of green glass with marigold iridescence.

PANELED SMOCKING

I've always felt this pattern was from either England or Australia, but I may be wrong on both count Marigold is the only color I've seen, and there are no other shapes reported.

PANELED TREETRUNK

While similar to the Northwood Treetrunk vase, this scarce and interesting vase has an appeal all its ow The example shown is 7½" tall and has a base diameter of 4⅞". It has eight panels, and the coloring is a fi amethyst. I suspect it was made by Dugan but can't be positive.

PANSY

The Pansy bowl shown typifies the Imperial quality so often found in iridized glass. The luster is outstandin with a gleaming finish equal to the best Millersburg we all treasure so much. The mold work is super, enhance by the very rich gold finish. While the Pansy pattern doesn't bring top dollar, it is a pleasure to own such beautiful item.

PANTHER

One of Fenton's favorite animal patterns, Panther is realistic and found in footed berry sets. The small bow in marigold are available, but the master bowls, especially in darker colors, are scarce. The example shown a rare green, but Panther is known in marigold, blue, green, amethyst, red and a very rare white.

PARLOR PANELS

If you haven't had a chance to see one of these beautiful 4" vases, especially the Imperial purple shown, you' certainly missed a real experience, for these are a glass collector's dream. Parlor Panels was shown in o Imperial catalogs and has not been reproduced to date. I've heard of marigold ones and I'm sure smoke and gree are possibilities. Some examples are swung to 12" lengths.

PASTEL PANELS

Usually found in pitchers and tumblers, we show a handled mug in this nicely done pattern. As you can se the coloring is a very strong ice green, but this pattern is available in other pastels as well. I suspect Imperi to be the maker but can't be sure.

PEACH

Once more we show a Northwood pattern that was also produced in other types of glass. This was more tha likely one of the earlier Northwood Carnival glass patterns and is found in a very fine cobalt blue as well a white, the latter often with gilting. What a shame more shapes were not made in Peach, for only berry sets, tab sets and water sets are known. The pattern is scarce and always brings top dollar.

PEACH AND PEAR

While no shards of this pattern were catalogued from the Dugan digs, I'm convinced it was their product. is available mostly in marigold wtih an occasional amethyst one turning up. The mold work is excellent and on the one shape and size are known.

Paneled Treetrunk

Pastel Panels

Paneled Holly

Pansy

Panther

Paneled Smocking

Parlor Panels

Peach

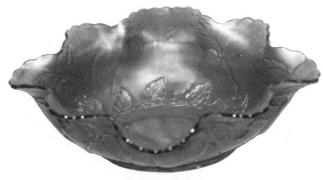

Peach and Pear

PEACOCK (MILLERSBURG)

While both Fenton and Northwood had similar patterns, the Millersburg Peacock is most noted and by far the best example of this design. The detailing is considerably greater, the bird more realistic and the quality of glass and finish superior. There are slight differences between the large bowl and the individual berry bowls in the Peacock pattern, but most of these are in the detail permitted on the allowed space and a complete berry set is worth searching for. The color most often seen is a fine fiery amethyst, but green, marigold and clambroth exist as well as blue.

PEACOCK AND DAHLIA

While this Fenton pattern is related in design to the Peacock and Grape, it is a better design and much scarcer. Known only in bowls and plates, Peacock and Dahlia carries the Berry and Leaf Circles pattern on the exterior. Colors known are marigold, blue, green, amethyst, vaseline, white and aqua.

PEACOCK AND GRAPES

Obviously, this Fenton design and Peacock Dahlia were from the same artist. This too, is found on bowls, flat or footed, and scarce plates. Colors are marigold, blue, green, amethyst, aqua, vaseline, white, red and peach opalescent.

PEACOCK AND URN (FENTON)

The Fenton version of this pattern (both Northwood and Millersburg had their own) is most often seen on stemmed compotes in marigold or aqua. Other shapes are bowls, plates and a scarce goblet from the compote mold. Colors known are marigold, blue, green, amethyst, white, aqua, lavender and red.

PEACOCK AND URN (MILLERSBURG)

Millersburg Peacock and urn differs from the regular Peacock pattern in that it has a bee by the bird's beak. The shapes known are large bowls or small 6½" bowl, a chop plate, and a giant compote in the usual colors of marigold, green and amethyst.

PEACOCK AND URN (NORTHWOOD)

This subject must have been very appealing to the mass market for not only did Northwood have a version of Peacock and Urn but so did Fenton and Millersburg. And while there is little chance to confuse the Fenton version with the others, the Millersburg and Northwood patterns are quite similar. The Northwood Peacock and Urn is most often found on ice cream sets, has three rows of beading on the urns, and has more open space between the design and the outer edges of the bowl. Colors include both vivids and pastels as well as aqua opalescent. Shown is the rare 6" plate in marigold.

PEACOCK AND URN "MYSTERY" BOWL (MILLERSBURG)

There is little mysterious about this bowl any longer. It is indeed from Millersburg, has a bee, two rows of beading on the urn and is found in both the ruffled and ice cream shape in 8" and 8½" diameters. Colors are the usual marigold, amethyst and green and all are rare.

PEACOCK AND URN VARIANT (MILLERSBURG)

Over the years, so much controversy has arisen over all the Millersburg Peacock designs, I'm showing all of them including this 6⅛" very shallow bowl. Note that there is a bee and the urn has three rows of beads! The only color reported on this piece is amethyst and so far, four are known so it is rather rare.

Peacock and Urn "Mystery" Bowl

Peacock and Urn Variant

Peacock

Peacock and Dahlia

Peacock and Grapes

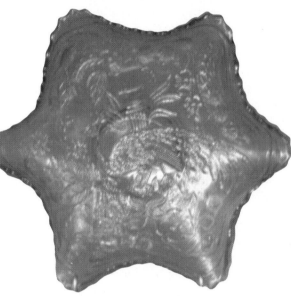

Peacock and Urn (Fenton)

Peacock and Urn (Millersburg)

Peacock and Urn (Northwood)

151

PEACOCK AT THE FOUNTAIN

Peacock at the Fountain probably rates as Northwood's second most popular pattern, right on the heels of th famous Northwood Grape, and it's so easy to see why. It is an impressive, well done pattern with an intriguin design. Available in berry sets, table sets, punch sets, water sets, compotes and a large footed orange bow Peacock at the Fountain was made in a host of colors, including the much-prized ice green and aqua opalescen Shards of this pattern were found at Dugan, leading us to believe Harry Northwood "farmed out" work to th Dugan Glass Company to keep up with demand on his best-selling patterns. Dugan made a copy cat water se to be sure.

PEACOCK GARDEN VASE

This beautiful, well iridized vase was an early product of Northwood Carnival Glass and when the Northwoo factory closed, many of the molds were purchased by the Fenton company – thus, the reason this vase was re issued by the Fenton company a few years back in the other types of glass. At any rate, the Northwood Peacoc Garden vase is truly a rare and beautiful sight to behold and a real treasure for the owners. The coloring i exceptional.

PEACOCK LAMP

This beautiful lamp base is on a crystal base glass, but with an enameled interior and an iridized exterio Colors are marigold, smoke, amethyst and red. Occasionally there is no hole in the bottom indicating it wa intended as a vase. Its height is 10¼", and sometimes it is found with a wooden base.

PEACOCK TAIL

Very reminiscent of the Northwood Nippon pattern, Fenton's Peacock Tail is known on bowls of all sizes, bor bons, compotes, hat shapes (some with advertising) and vase interiors. Colors I've seen are marigold, blue, greer amethyst, amber and red.

PEACOCK TAIL AND DAISY

I am very thrilled to finally show this very rare Fenton bowl pattern. Besides the marigold, there is a amethyst and a blue opal milk glass that isn't iridized, but I've heard of no other shapes or colors. As you ca see, the pattern is graceful as can be, and the design is very well balanced.

PEACOCK TAIL VARIANT

One look at the iridescence of this little compote, and it becomes quite obvious it is a Millersburg product culled from three common motifs used by most glass companies of the day – the peacock tail rings, the stipple rays and the feather. At any rate, it is a nice little item and quite enjoyable to own. Amethyst seems to be th most often seen color but marigold and green are known.

PEACOCK VARIANT (MILLERSBURG)

This extremely rare 7½" (same size as the Courthouse bowls) variant is an outstanding example of Millersbur craftsmanship. The mold work is very sharp with great detail and there is a bee, but no beading. Also the ur is quite different than the other peacocks. Colors are marigold, green and amethyst, but blue may exist.

Peacock Tail and Daisy

Peacock at the Fountain

Peacock Garden Vase

Peacock Lamp

Peacock Tail

Peacock Tail Variant

Peacock Variant (Millersburg)

PEACOCKS

Often called "Peacocks on the Fence," this Northwood pattern typifies what Carnival glass is really all abou[t] An interesting pattern, well molded, and turned out in a variety of appealing colors, Peacocks is a delight. On[ly] average size bowls and plates are known, but the color range is wide, including vivid colors, pastels and a real[ly] beautiful aqua opalescent. The exterior usually carries a wide rib pattern.

PEBBLE AND FAN

This very attractive vase is 11½" tall with a base diameter of 4½" and is truly a triumph. The mold work a[s] well as the basic design is quite good. Colors I've heard about are cobalt blue, marigold and a rather unusu[al] amber with vaseline finish.

PEBBLES (DUGAN)

Here is the Dugan version of Coin Dot and as you can see, it has no stippling. Found on open bowls and plate[s] colors are amethyst, marigold and green.

PENNY MATCH HOLDER

Once again we picture a rare but useful novelty that is seldom found. The octagonal base is 3⅜" across an[d] the entire Match Holder is 3½" tall. Purple is the only color I've heard of, and the iridescence is very rich an[d] heavy, very much like the better Northwood products in this color, but the maker remains unknown.

PEOPLE'S VASE

We are told this masterwork of art glass was produced by the Millersburg company to show appreciation fo[r] the help of the Amish people of the area in getting the factory started. The dancing figures represent Amis[h] children, hands clutching, prancing happily in thanks for a successful harvest – the only time their religio[n] permitted such merriment. In the background, Grandfather looks on with pleasure. But regardless of th[e] circumstances surrounding the vase's conception, there is no doubt that this is at the very top of Millersburg[s] best and surely deserves a place in the history of American glass. The only pattern variation occurs in the li[p] which is usually straight but very occasionally appears in scallops. The vase is 11½" high, 5½" in diameter an[d] weighs 5 lbs. The colors are marigold, amethyst, green and cobalt blue.

PERFECTION

This beautiful water set is a fitting companion to its sister design, the Gay 90's. Outstanding mold work, col[or] and iridescence are the hallmark of Millersburg and the Perfection pattern certainly fits this description. Th[e] pitcher is 9½" tall and quite bulbous. It has four mold marks. The tumblers are 4" tall and taper from 2⅞[] diameter at the top to 2⅛" at the base which has a 24-point rayed star.

PERSIAN GARDEN

Found in berry sets, ice cream sets and two sizes of plates, this very well designed pattern came from th[e] Dugan factory. It is available in marigold, amethyst, green, blue, white, ice blue and ice green.

PERSIAN MEDALLION

Very oriental in flavor, Persian Medallion is one of those well-known Fenton patterns that was extreme[ly] popular when it was made and is available on bowls of all shapes and sizes, plates both large and smal[l] compotes, rose bowls and even a hair receiver (not to mention interiors on punch sets). Colors are marigold, blu[e] green, amethyst, white, amber and red.

Pebbles (Dugan)

Persian Medallion

Peacocks

Pebble and Fan

Penny Match Holder

People's Vase

Perfection

Persian Garden

PETAL AND FAN

Found only in bowls of various sizes, this Dugan pattern also has the Jeweled Heart as an exterior design. The motif itself is simple but attractive – a series of alternating stippled and plain petals on a plain ribbed background with a fan-like design growing from each plain petal. The feeling is almost one of ancient Egypt where such fans were made of feathers. Petals and Fans is available in many colors including peach opalescent.

PETALS

This Dugan pattern is primarily a bowl pattern but is occasionally seen in a compote as well as a super banana bowl shape. Colors are the usual Dugan ones and the exterior seems to be found with a wide panel design.

PETER RABBIT

Peter Rabbit is one of Fenton's rare treasures, much sought by collectors and always priced for top dollar when sold. The shapes are 9" bowls and 10" plates in marigold, honey amber, green, blue and amethyst. The design is closely related to both Little Fishes and Coral patterns.

PICKLE PAPERWEIGHT

Of all the oddities I've come across in Carnival glass, this is probably the oddball of all time. It measures 4½" in length and is hollow. As you can see, it has a super color and finish. I assume it to be a paperweight for I can think of little else it could be used for. The maker is unknown, and the only color reported is amethyst.

PILLAR AND FLUTE (IMPERIAL)

Like so many good basic patterns from the Imperial company, this simple design of narrow flutes has great balance by echoing these same flutes on the base in a pretty mirror-image. The color is good strong marigold on the bowl only.

PILLOW AND SUNBURST

This pattern is credited to the Westmoreland company and while it has been seen on several shapes in crystal, the only reported iridized shape is the bowl shape. The pattern is exterior and can be found in marigold or purple. The example shown is an 8½" bowl.

PIN-UPS

This scarce Australian pattern, simple but distinctive, is found mostly on 8½" bowls in a rich purple, but here we show a very scarce marigold. The exterior carries a slender thread border.

PINCHED SWIRL

We seldom hear much about this attractive design, but it is known in both rose bowls and a rare spittoon, both in peach opalescent. I don't believe the maker has been confirmed, but Dugan or Westmoreland seem likely candidates.

PINE CONE

What a little beauty this design is! Known only on small bowls and scarce plates. Fenton's Pine Cone is seen mostly on marigold, blue or green, but an occasional amber plate has been seen. The design is well molded, geometrically sound, the the iridescence is usually quite good.

Pillar and Flute (Imperial)

Pickle Paperweight

Petals

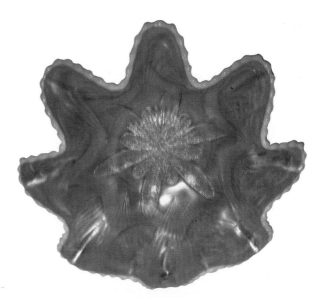

Petal and Fan

Peter Rabbit

**Pinched
Swirl**

Pillow and Sunburst

Pin-Ups

Pine Cone

PINEAPPLE

Sometimes called Pineapple and Bows, this Sowerby pattern is available in 7½" dome-footed bowls, a comp— and a creamer. The mold work is quite good with much attention paid to detail. The luster is very good on t— items I've seen and colors of marigold, purple and blue are known.

PINWHEEL VASES

Now known to be of English origin, these pretty vases are found in three sizes in colors of marigold, blue a— amethyst. The mold work and coloring are very good and so is the design. Originally this pattern was cal— Derby.

PIPE HUMIDOR

Here is a fantastic Millersburg pattern many collectors haven't even seen because it is just that rare. And wh— a pity so few of these are around since it is lovely enough to grace any collection. The coloring and iridescen— are exceptional and the design flawless. Imagine how difficult to remove the lid from the mold without dama— to the pipe and stem. The humidor is just over 8" tall and measures 5" across the lid. A three-pronged spon— holder is inside the lid, intended, of course, to keep the tobacco moist. Around the base is a wreath of acorns a— leaves above another leaf-like pattern that runs down the base.

PLAID

On the few occasions this Fenton pattern is found, the coloring is usually cobalt blue with heavy luster. T— example shown is a rare red. As you can see, the coloring is very bright with rich iridescence. Needless to sa— these 8½" bowls are a collector's dream and certaily would compliment any collection of red Carnival glass.

PLAIN JANE (IMPERIAL)

About all this Imperial bowl has going for it is the beautiful smoke finish and the radium luster, for t— interior has no pattern and the back has only a wide panel with a star base. Still it is somehow special.

PLAIN JANE BASKET

Long credited to the Imperial company, this large basket, while rather plain, has great appeal. It has be— seen in smoke as well as marigold and measures 9¼" to the top of the handle.

PLAIN PETALS

Here is a pattern seldom dicussed, yet it is found as the interior design of some pieces of Leaf and Beads. — is my opinion the latter was made by both Northwood and Dugan, and these pieces with the Plain Petals interi— as well as the famous rose bowl are Northwood. The example shown is on a very interesting leaf-shaped nap—

PLUME PANELS

For years I felt this rather stately vase was a Northwood product but as you can see, it has been found — a beautiful red so I'm rather sure it came from the Fenton factory. Other colors known are marigold, blue, gree— amethyst and white.

Plain Jane (Imperial)

Plain Petals

Pineapple

Pinwheel Vases

Pipe Humidor

Plaid Bowl

Plain Jane Basket

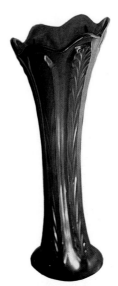

Plume Panels

POINSETTIA (IMPERIAL)

Found only on one shape (the beautiful milk pitcher shown), Poinsettia is an outstanding Imperial patter Standing 6½" tall, the Poinsettia is usually found in marigold or smoke color, but as you can see, a rare an beautiful purple does exist as does an equally rare green. What a shame more shapes do not exist in th beautiful design!

POINSETTIA (NORTHWOOD)

For some reason, in years past, someone attributed this very stylish pattern to the Fenton Glass Compan Just why, I can't guess, for Poinsettia was made by Northwood in custard glass and was illustrated in the advertising of the day. At any rate, this mistake has been corrected, and we now recognize this well done bo as a Harry Northwood design. Poinsettia is found either as a flat based or footed bowl with the Fine Rib as exterior pattern. The finish is nearly always superior. The colors are marigold, green, purple, fiery amethys white and ice green.

POINSETTIA (INTERIOR)

Here is something unusual – a tumbler with all the pattern on the inside. Of course, we've all seen th Northwood Swirl pattern which is also an interior one, but that was a simple geometrical design while th Interior Poinsettia is an offering of a large flower. Apparently these were never very popular for they are ve scarce. Also, to the best of my knowledge, no pitcher has been found. The iridescence is on both the inside an outside and is a good rich marigold. Only some of these Northwood tumblers are marked.

POND LILY

Much like other Fenton patterns such as Two Flowers and Water Lily, Pond Lily has both scale filler and th Lotus-like flower. The only shape I've seen is the bonbon, and colors reported are in marigold, blue, green an white. Of course, other colors may exist and certainly red is a possibility.

PONY

For years the origin of this attractive bowl has been questioned and only recently has it been attributed the Dugan company. Colors seen are marigold, amethyst and ice green. The mold work is quite good.

POPPY (MILLERSBURG)

Large open compotes seem to have a fascination all their own, and this one from Millersburg is certainly exception. It is quality all the way, whether found in green, purple, or marigold. Standing 7" tall and being across, Poppy has four mold marks. It has the Potpourri as a secondary pattern. The poppy flowers and leav are well done and are stippled.

**Poinsettia
(Imperial)**

**Poinsettia
(Northwood)**

Poinsettia (Interior)

Pond Lily

Pony

Poppy

POPPY (NORTHWOOD)

This Northwood pattern is most often found on small oval bowls, described as trays or, with the sides crimp as pickle dishes. However, it is also found as an exterior pattern on larger bowls, some with plain interiors, oth with a large Daisy in the center of the bowl. The colors are electric blue, marigold, peach, purple, aqua opal, a white. Others may exist but these are the ones I've seen.

POPPY SHOW (NORTHWOOD)

Let me state from the beginning this is not the same pattern as the Poppy Show Vase. It is a beautiful w made item, very much akin to the Rose Show pattern in concept and design. It is found only on large bowls a plates in a wide range of colors, including marigold, green, blue, purple, white, ice blue and ice green. T Northwood pattern brings top dollar whenever offered for sale.

POPPY SHOW VASE (IMPERIAL)

What a shame this beauty was chosen to be reproduced by Imperial in the 1960's. The old Poppy Show V is a real show stopper, standing about 12" tall, with a lip diameter of 6¾"! The mold work is very fine, with t graceful poppy in a series of four panels around the vase. I've seen this artistic gem in marigold, clambroth, pas marigold, amber, green and purple. Naturally, the darker colors are quite scarce and are priced accordingly

POTPOURRI (MILLERSBURG)

I certainly welcome the chance to show this very rare pitcher in the Potpourri pattern, a kissing cousin Millersburg's Country Kitchen pattern. Only two of these pitchers have been reported and both are marigo smaller than a water pitcher and were used for milk.

PRAYER RUG

Known only in the finish shown, beautiful custard glass with a marigold iridescence, Fenton's Prayer Rug a seldom-seen item. The only shape is a handled bonbon, but I suspect time will bring to light additional or since uniridized pieces are known in small bowls, vases and hat shapes.

PREMIUM

Not only found in the well-known candlesticks shown, but also in 8½" bowls, 12" bowls and 14" plat Imperial's Premium pattern is shown in old catalogs in marigold, clambroth, purple, green and smoke. T candlesticks are 8½" tall, heavy and beautifully iridized! While not in the class with the Grape and Ca camdlesticks, they are still quite nice. Used with the medium size bowl, they make a nice console set.

PRETTY PANELS

This is a marked Northwood tumbler and as such is quite a sight for tumbler collectors. The color is a ve bold frosty green, and the enameled cherry design is above average. It is found in marigold also.

Potpourri (Millersburg)

Poppy

Poppy Show

Poppy Show Vase

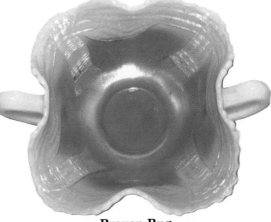

Prayer Rug

Premium

Pretty Panels

PRIMROSE

For some reason, this Millersburg pattern hasn't received the "raves" it might, and I can't understand wh
I have one and treasure it just as much as I do a Nesting Swan. The glass work is equally as good and tl
iridescence excellent. The reverse pattern is Fine Cut Hearts and the two blend beautifully. Primrose is four
primarily on large bowls in the three main Millersburg colors, plus the spectacular blue shown.

PRIMROSE AND FISHNET (#2475)

This unusual Imperial pattern has two kissing cousins, also in red Carnival; one showing grapes, the oth
roses. The floral design is on one side only, and the fishnet covers the remainder of the glass. While red is tl
only reported color in iridized glass, all three patterns are known in crystal. The Primrose and Fishnet va.
stands 6" tall. Needless to say, they are quite scarce.

PRISM AND CANE

This very scarce product of Sowerby has an interior pattern of Embossed Scroll variant and, as you can se
the base is ground. Apparently it is a sauce dish or jam dish as it measures 5" across the rim and stands 2l.
tall.

PRISM AND DAISY BAND

Apparently one of Imperial's late designs in Carnival glass and intended for a cheap mass sale, Prism ar
Daisy Band can be found only in marigold in berry sets, breakfast sets, a stemmed compote and a vase shap
The coloring is adequate but not superior.

PRISMS

We now have evidence to support a Westmoreland origin for this unusual little compote. For quite awhil
amethyst was the only color seen, but here is a marigold of which I've seen some four or five, and green als
exists. prisms measure 7¼" across the handles and is 2½" tall. The pattern is all exterior and is intaglio wit
an ornate star under the base like the one on the Cherry and Cable butter dish.

PROPELLER

Besides the usual small compote found in this Imperial pattern, I'm very happy to show the rare 7½" stemme
vase in the Propeller pattern. The coloring is a good rich marigold, but others may exist since the compote is see
in marigold, green and amethyst.

Primrose

Primrose and Fishnet

Prism and Cane

Prism and Daisy Band

Prisms

Propeller

PULLED HUSK CORN VASE

Apparently Harry Northwood wasn't quite satisfied with this very rare example of a corn vase, for few of these are around in comparison with the regular corn vase. Known in two sizes, the Pulled Husk vase has been seen in green and purple and some are pulled more grotesquely than others.

PULLED LOOP

This rather simple Fenton vase design is found quite often, mostly in marigold, blue or amethyst, but it known in a beautiful green, as well as an occasional peach opalescent finish. The size may vary from 8" to 12 but the finish is usually very heavily lustered.

PUZZLE

Found in stemmed bonbons and compotes, Dugan's Puzzle is an appealing pattern. The all-over design is we balanced and the stippling adds interest. Colors known are marigold, purple, green, blue, white and peac opalescent.

QUEEN'S LAMP

I first shwowed this very rare lamp in my Millersburg book, and I still believe it came from that compan. There are two green iridized ones known as well as a crystal example with a matching shade. The base of the lamp measures 7" across and the lamp to the top of the font is 9½" tall.

QUESTION MARKS

Here is a simple Dugan pattern found on the interiors of bonbons and occasionally compotes like the or shown. Again the exterior is usually plain, and the colors are peach opalescent, marigold, purple and white. Bot the compote and the bonbon are footed; the compote is one of the small size, measuring 4½" tall and 4" acros the highly ruffled edge. The exterior occasionally has a pattern called Georgia Belle.

QUILL

Once again we show a pattern of which shards were found in the Dugan diggings, and I truly believe Qui was indeed a Dugan Glass Company pattern. The pitcher is some 10" tall and has a base diameter of 4½". Th colors are marigold and amethyst, and the iridescence is usually above average. Quill is a scarce pattern an apparently small quantities were made, again pointing toward the Dugan company as the manufacturer. Th water set is the only shape.

QUILTED DIAMOND

Certainly one of Imperial's best exterior designs, this pattern is found as a companion to the Pansy patter on one-handled nappies, dresser trays and pickle dishes. Just why Imperial didn't use it more is a mystery fo it is beautifully done and very pleasing to the eye.

RAGGED ROBIN

Just why more of these Fenton bowls weren't produced is a mystery, but the fact remains these are quite har to find today. Found only on average size bowls often with a ribbon candy edge. Colors reported are amethys marigold, blue, green and white with blue most available.

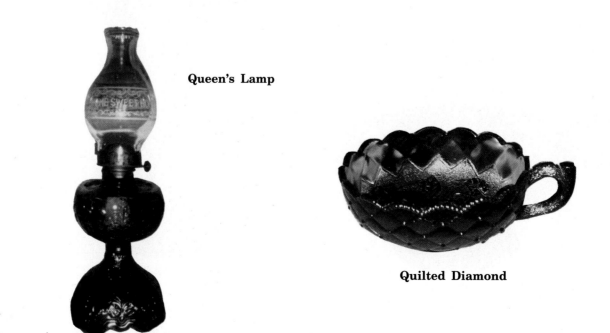

Queen's Lamp

Quilted Diamond

Pulled Husk Corn Vase

Pulled Loop

Puzzle

Question Marks

Quill

Ragged Robin

RAINBOW (NORTHWOOD)

Similar to the Northwood Raspberry compote or the one with only a Basketweave exterior, this one completely plain with iridescence on the inside only. The luster swirls around the glass in layers, just like rainbow, thus the name.

RAINDROPS

Here is another of the dome-footed bowls available in peach opalescent like so many offered by the Dug company. Remember, I said earlier that I felt Dugan was responsible for at least 90% of the peach Carnival, a a close study of these bowls will support this belief. Raindrops is round without stippling. It has the Keyh pattern as an exterior companion and has four mold marks. All in all, it is a nice pattern to own, especially you like the peach opalescent glass. Also found in amethyst.

RAMBLER ROSE

Until quite recently, I'd always felt Rambler Rose was a Fenton product, but upon examining a large sha of this pattern from the Dugan dump site, I'm compelled to admit my mistake. This water set has a bulbo pitcher with a ruffled top. The flowers are well designed and clearly molded. The colors are marigold, purple, bl and green. Perhaps research in the years ahead will add more information about this pattern.

RANGER

Known in breakfast sets, water sets, a milk pitcher and table sets, this pattern, often confused with t Australian Blocks and Arches pattern, is found only in marigold. It is an Imperial product, but one version the tumbler is known to have been produced by Christales de Mexico and is so marked.

RASPBERRY

Even without the famous trademark, this pattern would be recognized as a Northwood product, for it includ the basketweave so often found on that company's designs. Available in water sets, table sets, berry sets and milk pitcher. Raspberry has long been a favorite with collectors. The colors are marigold, green, purple, ice bl and ice green with the richly lustered purple most prevalent.

RAYS

If you look closely at this Dugan pattern you will detect an inner ring of pointed rays before the veins f out toward the outer edges, distinguishing the simple design from similar ones. It was made in all the Duga colors and this one has a Jeweled Heart exterior.

RAYS AND RIBBON

Each of the makers of Carnival glass seems to have had a try at a pattern using stippled rays. The Millersbu version is quite distinctive because of the bordering of ribbon-like design, resembling a fleur-de-lis design. Mo of the bowls are not radium finish and usually carry the Cactus pattern on the exterior. Occasionally a plate found in Rays and Ribbons, but one wonders if this were not produced as a shallow bowl. Amethyst is the usu color, followed by green, marigold, and vaseline in that order.

RIBBON TIE

Sometimes called Comet, this well-known Fenton pattern is found chiefly on all sorts of bowls as well ruffled plates. The colors range from very good to poor in marigold, blue, amethyst, red and green, and often t luster is only so-so.

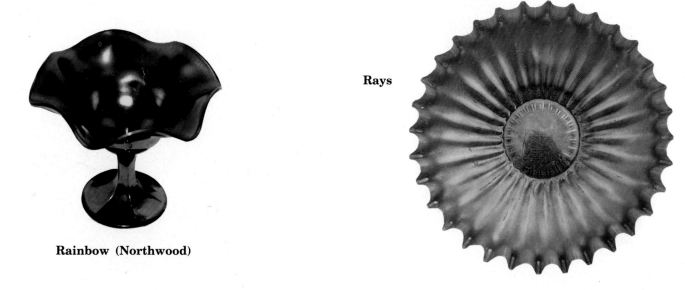

Rays

Rainbow (Northwood)

Raindrops

Rambler Rose

Ranger

Raspberry

Rays and Ribbons

Ribbon Tie

RINGS

This large vase (8" tall) isn't much on design but it certainly has a place on the practical side for holding larg floral displays. The coloring is a good rich marigold and the workmanship top notch.

RIPPLE

This Imperial pattern is fairly common in marigold, clear or amethyst glass but the teal color shown in a re exception and quite a beauty. Ranging in height from 10" to 16", the Ripple vase depends on what design it ha in the "pulling" or "slinging" of the glass while hot and, naturally, the taller the vase, the less design exists

RISING SUN

This very unusual pattern has been seen in both the marigold shown and cobalt blue. The pitcher is foun in two variations, one with a pedestal base, the other flat base. A matching tray has been reported but I haven seen it. The maker is U.S. Glass. A rare table set was made in marigold. Shown is the covered sugar.

ROBIN

Despite the reproductions of this fine old Imperial pattern, the prices have held up rather well on the ol pieces. The water sets, found only in marigold, and the mug found in smoke are escpecially desirable. Apparentl the appeal lies in the handsome presentation of the nicely done bird, the flower and branch dividers, and th flowering leaf pattern, so pleasing to the eye.

ROCOCO

This beautiful little Imperial vase was the first item in smoke I'd ever seen, and I must admit I loved it a first sight! While it may be found on a small bowl shape with a dome base and in marigold as well as smok it is the vase most think of whenever Rococo is mentioned. The vase is 5½" tall and shows four mold marks. have had a green one reported but haven't seen it yet.

ROMAN ROSETTE GOBLET (UNLISTED)

While this pattern is not difficult to find in pressed glass, this is the only reported item in iridized glass t the best of my knowledge. As you can see, it is slightly crooked, but the iridization on the clear glass unmistakable with beautiful blue and pink coloring. It measures 6" tall and has a base diameter of 2¾".

ROSALIND (MILLERSBURG)

Often called Drape and Tie, this classic Millersburg design is found as the interior pattern on Dolphi compotes, on rare 9" compotes and large and small bowls. The colors are marigold, amethyst and green, but I'v seen a 9" bowl in a stunning aqua and a plate is known in green.

ROSALIND VARIANT (MILLERSBURG)

If you look closely at this design and the regular Rosalind shown elsewhere, you will see a good deal difference, especially in concentric peacock tail sections. For this reason, I'm listing the compote shown as variant (and expect letters about it). Colors reported in this compote are amethyst and green but marigold wa probably made also.

Rings

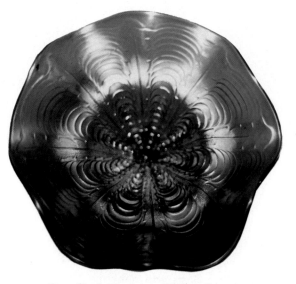

Rosalind Variant (Millersburg)

Ripple

Rising Sun

Robin

Rococo

Roman Rosette Goblet

Rosalind

ROSE AND GREEK KEY PLATE

This very beautiful square plate is a sight to behold. Not only is it quite unique but so very well designed th I simply cannot understand why there aren't more of these. But alas, there's only the one known. The colori is a smoky amber. The plate measures 8½" across, and the roses are deep and hollow on the underside, mu like the well-known Rose Show bowls.

ROSE COLUMN

This stately Millersburg vase is a real beauty and like several Carnival patterns such as the Imperial Gra carafe or the Grape Arbor pitcher, the rows of roses are hollow. This requires great skill by the worker wh removing it from the mold, and many must have been broken in doing so. This lovely vase is 10" tall and across the diameter. There are six columns of roses, each topped by a spring of leaves. The iridescence is t quality, and the colors are marigold, blue, green and amethyst. The Rose Column vase would make a lov companion to the People's Vase.

ROSE GARDEN

Once again we finally have a true maker for a much-disputed pattern, and Rose Garden is now in the list glass from the Eda Glassworks of Sweden. It is found in many shapes including bowls, round and oblong vas a pitcher that is rare, a beautiful rose bowl and a covered butter dish. Colors are marigold and blue only.

ROSE SHOW

What a handsome piece of glass this is! The design of this Northwood is flawless, heavy and covering eve inch of available space. Yet it isn't in the least bit busy looking. One has the distinct feeling he is looking i a reeded basket of fresh-cut roses and can almost smell the perfume. Found only on bowls and a plate, t beautiful pattern was produced in small amounts in marigold, purple, blue, green, white, ice blue, ice gree peach opalescent, amber, aqua opalescent and a rare ice green opalescent.

ROSE SHOW VARIANT

Over the years, very little has been learned about this plate variant of the Rose Show pattern, but as you c see, it is quite different in both flower and leaf arrangement and number. It is found in a host of colors includi marigold, amethyst, green, blue, peach opalescent, aqua opalescent, ice blue and white.

ROSE SPRAY COMPOTE

Standing only 4½" at its tallest point, this beautiful compote is a real treasure. The only colors I've heard abo are white and the beautiful ice blue and ice green. I suspect this is a Fenton product, but have no verificati of this. The rose and leaf spray is on one side of the rim only and is rather faint, much like the Kittens bo

ROSE TREE

Make no mistake about it, this is a very scarce and desirable Fenton bowl pattern. I believe this was t Fenton answer to the Imperial Lustre Rose, but apparently it was made in small quantities. The colors kno are marigold and cobalt blue, and the size is a generous 10" diameter.

Rose Show Variant

Rose and Greek Key Plate

Rose Column

Rose Garden

Rose Show

Rose Spray Compote

Rose Tree

ROSES AND FRUIT

This beautiful little compote is unique in several ways and is a very hard-to-find Millersburg item. It measu 5¼" from handle to handle and is nearly 4" high. Notice the deep bowl effect (so often used by Millersburg compotes and bonbons) and the pedestal base. In addition, observe the unusual stippling around the edge of interior, and, of course, the combination of roses, berries, and pears are quite distinctive. It is found in green a amethyst mostly, but was made in marigold and blue too.

ROSES AND RUFFLES LAMP (RED)

I'm frankly not too taken with Gone With the Wind Lamps, but the beautifully iridized ones are in a cl by themselves, and the very few ones known are simply beautiful. The Roses and Ruffles lamp is 22" tall. It I excellent fittings of brass and is quality all the way. The mold work on the glass is quite beautiful and the lus superior.

ROSETTE

Combining several well-known Carnival glass patterns, including Stippled Rays, Beads and Prisms, t Northwood pattern isn't the easiest thing to find. In arrangement, it reminds one of the Greek Key patterns, I Rosette stands on its own. Found only on generous sized bowls, the colors are marigold and amethyst. Green m be a possibility but I haven't seen one.

ROUND-UP

As I stated earlier, Round-Up, Fanciful and Apple Blossoms Twigs all have the same exterior pattern a shards of the latter two were found in the Dugan diggings. Found only on bowls, ruffled plates and true plat Round-Up is available in marigold, purple, peach opalescent, blue, amber, white and a pale shade of lavend The true plate is quite scarce and always brings top dollar.

ROYALTY

Perhaps one might mistake this for the Fashion pattern at first glance but with a little concentrated stu it becomes obvious they aren't the same. Royalty is a pattern formed from a series of hobstars above a series diamond panels. Found on punch sets and two-piece fruit bowls, this Imperial pattern is mostly found marigold, but the fruit bowl set has been reported in smoke.

RUFFLED RIB SPITTOON

This little Northwood spittoon could be called Fine Rib or Lustre and Clear I guess, but I feel the name giv is more appropriate. At any rate, the coloring is a good rich marigold and the ribbing is on the interior. T spittoon stands 4" tall and has a rim diameter of 4½" with a collar base diameter of nearly 2".

Roses and Fruit

Roses and Ruffles Lamp

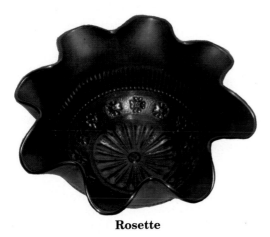

Rosette

Round-Up

Royalty

Ruffled Rib Spittoon

RUSTIC

I would venture to say Rustic is one of the best known and most common of vase shapes for it is plentif[ul] found in many sizes and colors including marigold, blue, green, amethyst, aqua, vaseline, amber, white, pea[ch] opalescent and red. It was pulled from the Fenton Hobnail vase shapes and can be found in sizes from 6" to 2[0"].

S-REPEAT

Made in crystal, decorated crystal, and gilt glass prior to being made in Carnival glass, S-Repeat is found [in] only a small range of shapes in iridized glass. Besides the punch set shown, there is a rare toothpick that h[as] been widely reproduced and a handful of marigold tumblers that some believe are questionable. At any rate, [in] Carnival glass, Dugan's S-Repeat is a very scarce item.

SACIC ASHTRAY

This little ashtray is a real mystery in many ways. It reads: "NARAJA SACIC POMELO." Apparently it w[as] meant to go to Brazil, but why an English glass maker would mold such an item escapes me. The color is qu[ite] good, with a touch of amber in marigold.

SAILBOAT

Found in small bowls, plates, compotes, goblets and wines. Sailboats is a well-known Fenton pattern th[at] competes nicely with Imperial's Windmill pattern. Colors I've heard about are marigold, blue, green, vaselin[e,] amber and red, but not all colors are found in all shapes.

SCALE BAND

While not too original, this Fenton pattern of smooth rays and bands of scale filler does quite nicely in t[he] shapes chosen: bowls, plates, pitchers and tumblers. The color most seen is, of course, marigold. However, Sca[le] Band can also be found in green, amethyst, peach opalescent, aqua opalescent (quite rare) and red.

FISH-SCALES AND BEADS

Most of these bowls I've seen are small – 6" to 7" in diameter. Nevertheless, they are well done, interest[ing] items and add much to any collection. The Fishscale pattern is on the interior while the Beads design is on t[he] exterior. When held to the light, one pattern fits happily into position to complement the whole, giving a pleasa[nt] experience.

SCARAB HATPIN

I have always avoided showing hatpins or buttons but the one here is so pretty, I couldn't resist. It is on [a] very deep amethyst base color and has super luster. It measures 1¾" x 1⅜".

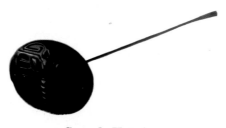

Scarab Hatpin

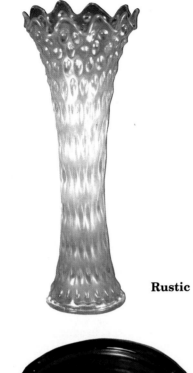

Rustic

S-Repeat

Sacic Ashtray

Sailboats

Scale Band

Fish Scale and Beads

SCOTCH THISTLE

What a pretty pattern this is for the interior of a compote! As you can see, the exterior is plain like so ma
compotes but the edges have a very interesting ruffled effect. Colors I've seen are marigold, blue, green a
amethyst but others may certainly exist. This pattern was manufactured by Fenton.

SCOTTIE

Surely there isn't a collector that hasn't seen these in the flea markets in Scotties, poodles, deer, duckies
rabbits. They are all covered powder jars, all marigold and made near the end of the Carnival glass heyday. S
the coloring is usually respectable, and they are cute little critters.

SCROLL AND FLOWER PANELS

I've always been intrigued by this stylish Imperial pattern even though it was reproduced in the 1960's w
a flared top. As you can see, the mold work, while very busy, is quite satisfying and the coloring is super. I
seen this vase in marigold and purple, but green is a possibility. The vase stands 10" tall on a collar base w
a many-rayed star.

SCROLL AND GRAPE (MILLERSBURG)

Nothing pleases me more than showing an unlisted pattern, especially if it is Millersburg! The only exam
reported of this Multi-Fruits and Flowers punch bowl interior is such a find. From the center, four strings
acanthus leaves extend to four clusters of grapes, all topped by a rich scroll. What a shame this wasn't a standa
pattern! It's as pretty as anything Millersburg ever made.

SCROLL EMBOSSED (IMPERIAL)

As I said earlier, while this pattern originated at the Imperial factory, it was later produced in England
Carnival glass, notably on the Diving Dolphins bowl and on a four-handled sauce dish. The Imperial version
used with File pattern exterior on bowls, with Easter Star as an exterior pattern on large compotes, as well
alone on bowls and smaller compotes. The usual colors are green, marigold or purple, but smoke does exist.
scarce plate shape is shown.

SCROLL EMBOSSED VARIANT (ENGLISH)

Much like its Imperial counterpart, this English version is found on small bowls, small handled ashtra
compotes and the Diving Dolphins bowls. Without the exterior design, it would be impossible to say who ma
which. However, the English version is known in marigold, blue, green and amethyst.

SEA GULLS BOWL

If one rarity in this book stands as an example of "scarce but not prized," the Sea Gulls bowl is that rari
Certainly there are far less of these to be found than many items that bring 10 times the money, but for so
strange reason, these cuties are not sought by most collectors. The two bird figures are heavily detailed as is t
bowl pattern. The color, while not outstanding, is good and is iridized both inside and out. The diameter of t
bowl is 5¾" and the depth is 2⅞". I believe the manufacturer was Jeanette but I could be wrong.

SEA GULLS VASE

This rare vase has been traced to the Eda Glassworks of Sweden, adding to their growing importance as a no
American maker of iridized glas. The only color reported is a rich marigold with a good deal of amber hue.

Scroll and Grape (Millersburg)

Interior of Scroll and Grape

178

Scotch Thistle

Scottie

Scroll and Flower Panels

Scroll Embossed

Scroll Embossed Variant

Sea Gulls Bowl

Sea Gulls Vase

SEACOAST PINTRAY

Pintrays are not common in Carnival glass, and this one from Millersburg is one of the nicest ones. Th irregular shaping, the beautiful coloring and fine detail make this an outstanding item. It measures 5½" x 3 and rests on an oval collar base. The colors are marigold, green, amethyst and a fine deep purple.

SEAWEED

Can you imagine a more graceful pattern that this one from Millersburg? The curving leaves and snail-lil figures seem to be drifting back and forth in the watery depths and the bubbles of beading add just the rigl touch. Usually found on fairly large bowls, Seaweed has three mold marks. The colors are marigold, green al amethyst. Also a rare plate and a rare small bowl have been found.

SEAWEED LAMP

I've heard of four of these lamps in two different base shapes, but all with the Seaweed design circling th body. All were in marigold ranging from quite good to poor. The example shown measures 12" to the top of th font. The maker is unknown but the lamp is a rare one.

SHELL

I've always felt this was a superior Imperial pattern – simple yet effective. This is especially true on the fe plates I've seen. The pattern is a well-balanced one of eight scalloped shells around a star-like pattern. Th background may or may not be stippled – I've seen it both ways. The shapes are smallish bowls, plates and reported compote I haven't seen. The colors are marigold, green, purple, smoke and amber.

SHELL AND JEWEL

Easily found, this Westmoreland pattern was made only in the shapes shown in colors of marigold, amethy and green (white has been reported but not confirmed). The pattern is a copy of the Nugget Ware pattern of th Dominion Glass Company of Canada. Shell and Jewel has been reproduced, so buy with caution.

SHRINE CHAMPAGNE (ROCHESTER)

Shown is one of three known stemmed champagnes manufactured by the U.S. Glass Company, to be give away at Shrine conventions (a rare toothpick holder is also known). The 1911 Rochester, New York champagr has painted scences of Rochester and Pittsburgh with a gilt decoration. The other champagnes are the 1910 Ne Orleans and Tobacco Leaf (Louisville, Ky. – 1909). Each is a premium example of the glass maker's skill.

SHRINE NIAGARA FALLS

Every so often a little jewel comes to light, like this toothpick holder. As you can see, it is a fine cobalt blu One side has the Shrine emblem and the other says: "Brown, Nagle, Willock, McComb – Niagara Falls." Th shape is that of a barrel giving added meaning and a touch of whimsey. Age is questionable so beware of inflate prices.

SILVER QUEEN

While this Fenton pattern has never been one of my favorite enameled water sets, it is certainly not all tha easy to find and has good color and finish. It is found only in marigold with the wide silver band and white scro and only in water sets.

Shrine Niagara Falls

Seacoast Pintray

Seaweed

Seaweed Lamp

Silver Queen

Shell

Shrine Champagne (Rochester)

Shell and Jewel

SINGING BIRDS

Found in custard glass, crystal and Carnival glass, Singing Birds is one of the better known Northwood patterns. The shapes are berry sets, table sets, water sets and mugs in vivid and pastel colors.

SINGLE FLOWER FRAMED

Much like the Single Flower pattern, this one has odd framing of double wavy lines that encircle the flower. The pattern is exterior, the colors the usual Dugan ones. Shapes are large and small bowls and the interior of the one shown shows a Fine Rib pattern.

SIX PETALS

Nearly always seen in shallow bowls in peach opalescent, Six Petals is occasionally found in a rare plate. The other colors available are purple, green, blue and white. A Dugan product. Shown in black amethyst in exterior view.

SIX-SIDED CANDLESTICKS

What a beauty this candlestick is! Not only is the near-cut design quite imaginative, but the workmanship just super. The Six-Sided Candlestick is 7½" tall and has a base diameter of 3¾". It has been seen in a rich marigold, an outstanding purple, smoke and green. It has been reproduced by Imperial in crystal, so be cautious in your selection!

SKI STAR

While this Dugan pattern is found occasionally on small bowls in purple, blue and green, it is on the large dome-footed bowls in peach opalescent where the pattern has its "day in the sun." These are found in many variations, some crimpled with one side turned down, others with both sides pulled up to resemble a banana bowl. The exterior usually carries the Compass pattern, an interesting geometric design, or is plain.

SLIPPER (LADY'S)

Here's another novel miniature, so ornamental yet appealing. Needless to say, very few of these were ever iridized so all are rare. The Lady's Slipper is 4½" long, 2¾" high, and is a good marigold. The entire piece is covered with stippling. From U.S. Glass.

SMALL RIB

The compote shown has been pulled into the rose bowl shape but versions of the lip opened out are also known. Coloring is much like the Daisy Squares pieces and may have the same common maker.

Single Flower Framed

Singing Birds

Six Petals

Six Petals

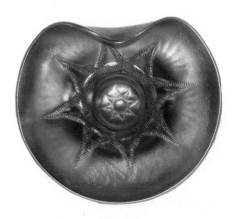

Ski Star

Slipper (Lady's)

Small Rib

183

SMOOTH PANELS

While this Imperial design is very much like the Flute #3 pattern, there are small differences. The mc obvious one is that the Flute panels are concave while the Smooth Panel ones are convex. Found in water se in marigold, green and smoke, or an 11" vase of pearl opalescent. The same vase has been seen in marigol amber, smoke, green and purple. The example shown has the Imperial "iron cross" mark.

SMOOTH RAYS (WESTMORELAND)

Like most of the other major manufacturers, Westmoreland had a try at a Smooth Rays pattern in compot and various size bowls with both flat and dome bases. Here is one of theirs in a beautiful blue milk glass iridize but these were made in marigold, amethyst, green, marigold milk glass, amber and teal as well.

SNOW FANCY

Known in small bowls as well as a creamer and sugar breakfast set, this scarce near-cut pattern is one n many collectors are familiar with. The bowl has been seen in green as well as a frosty white, and the breakfa set is shown in marigold; other colors may exist but I haven't seen them. It is a McKee product.

SODA GOLD

Much confused with Tree of Life and Crackle shown elsewhere in this book, Imperial's Soda Gold differs fro either in that the veins are much more pronounced and are highly raised from the stippled surface. It is fou only on short candlesticks, a rare 9" bowl, and beautiful water sets in marigold or smoke.

SODA GOLD SPEARS

Similar to Tree of Life and Crackle patterns, this all-over design is easy on the eye and a good backgrou to hold iridescence well. The bowl shown is the 8" size but 5" bowls and plates are known in both marigold a clambroth.

SOLDIERS AND SAILORS (ILL.)

One of two known such commemorative plates, this is the Illinois version. It features the Soldiers and Sailo Home in Quincy, Ill. and measures 7½". On the exterior is found Fenton's Berry and Leaf Circle design. T colors known are marigold, amethyst and blue.

SOUTACHE PLATE

I've long believed this to be a Northwood pattern, and while I've seen it on both bowls and lamp shades, th is the first footed plate I've run across. It measures 8¾" in diameter and has peach opalescent edging. The fo is a dome base and is of clear glass.

SPIRAL CANDLE

Like the Premium Candlesticks also made by Imperial, these heavy, practical candlesticks were sold in pai and were made in green, smoke and marigold. They measure 8¼" from the base to the top and as far as I kno did not have a console bowl to match. The smoke coloring is particularly beautiful with many fiery highlight

SPIRALEX

This is the name used in England to describe these lovely vases so I've continued using it. The colors a marigold, amethyst, green and blue, and all I've seen are outstanding with very rich iridescence. Sizes range fro 8" to 14".

Smooth Rays (Westmoreland)

Soda Gold Spears

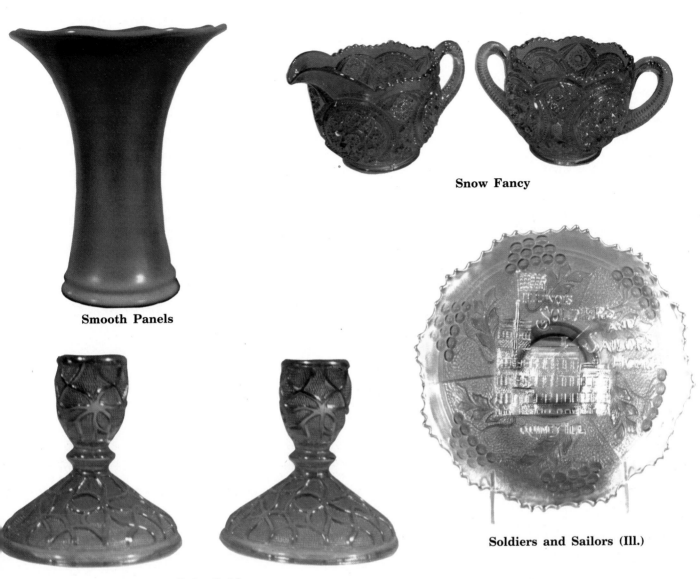

Smooth Panels

Snow Fancy

Soldiers and Sailors (Ill.)

Soda Gold

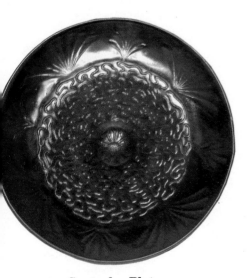

Soutache Plate

Spiral Candle

Spiralex

SPLIT DIAMOND

I'm very happy to show the complete table set in this Davison pattern, for while the creamer is easily foun the covered butter and sugar are quite scarce in this country. The only color is a good strong marigold with fi luster and superior mold.

SPOKES

Similar to the Corinth patterns, this Fostoria bowl is 10" across and is often seen in crystal. This if the fir iridized one I've seen and it has outstanding pink and blue highlights.

SPRING OPENING

Made from the standard Millersburg small berry bowl mold, this interesting hand-grip advertising plate is seldom-seen, always attractive item from the Ohio company. It measures 6½" in diameter, has the typical man rayed base and is known only in amethyst. The lettering says: "Campbell & Beesley Co. Spring Opening 1911

SPRINGTIME

In many ways, this pattern is similar to Northwood's Raspberry pattern, especially since both are bordered the bottom with versions of a basketweave. Springtime, however, is really its own master and bears panels wheat, flowers and butterflies above and throughout the basketweave. Found in berry sets, table sets and ve scarce water sets in marigold, green, amethyst and also in pastels, Springtime is a very desirable pattern.

SQUARE DIAMOND (COLUMN FLOWER)

Sometime after I gave this vase the above name, another author called it Column Flower so you have a choic It was made at the Riihimaki Glassworks in Finland and has been seen only in blue.

STAG AND HOLLY

This is probably the best known animal pattern in all Carnival glass other than the Peacock, and certain it remains one of Fenton's best efforts. Often brought out at Christmas time, the Stag and Holly is found main on footed bowls, rare footed plates and rare rose bowls. Colors are marigold, blue, green, amethyst, aqua, pea opalescent and red.

STAR

While there may not be a great deal of inventiveness to this English pattern, it certainly serves a usef purpose and does it with a good deal of attractiveness. The clear center shows off the star on the base quite we and the two tiny rows of rope edging add a bit of interest. Marigold is the only color I've heard about.

STAR AND FAN CORDIAL SET

This pattern is not the same as the vase with the same name, but has a close tie with the other cordial se like Zipper Stitch which were made in Czechoslovakia. It is a rare grouping, the only complete set known.

STAR AND FAN VASE

I've been privileged to see the two Imperial marigold vases shown, but a beautiful cobalt blue example al exists, and it must be something, for the marigold ones are very beautiful examples of the glassmaker's art. T glass itself is thick and very clear. The luster is rich and even, and the design is flawless. Star and Fan 9½" tall.

Spokes

Spring Opening

Square Diamond (Column Flower)

Star and Fan Cordial Set

Split Diamond

Springtime

Stag and Holly

Star and Fan Vase

Star

STAR AND FILE

Very much like Star Medallion in concept, Imperial's Star and File doesn't have the cane effect, but u panels of file, hobstars and sunbursts in a well-balanced design. Found in bowls, breakfast sets, a water set wine decanter set, handled vases, a rare rose bowl and a large compote, Star and File is known in marigo smoke, purple and green.

STAR AND HOBS

This very attractive 9" rose bowl has much to offer in both quality and looks. The color is a deep purple gla with beautiful gold luster and while the base pattern resembles Ohio Star, I believe this to be a Europe product.

STAR FISH

As you can see, this little compote is a real cutie. The design is strong and interesting, and the quality is t notch. Star Fish is from the Dugan company and is seen in marigold, green, purple and the beautiful pea opalescent.

STAR MEDALLION

Star Medallion is a much-overlooked, but well designed near-cut Imperial pattern. Apparently, it was a v popular pattern in its day for it is found in many shapes, including bowls of all shapes, a table set, a hand celery, a 9" plate, milk pitcher, punch cup, tumbler, goblet and a very beautiful compote. Colors are a ri marigold, smoke and occasionally a beautiful helios green.

STAR OF DAVID

For some unknown reason, only small amounts of this beautiful Imperial pattern must have been produc for it is seldom seen today. As you can see, it is simply a Star of David half-stippled on a plain background wi smooth ribbing to the edge of the bowl. The example shown measures 8¾" in diameter and is the famous hel green of Imperial with a silver finish, but the pattern is also known in marigold, smoke and purple. The exter carries the Arcs pattern.

STAR OF DAVID AND BOWS

The Northwood version of this figure (Imperial also produced a Star of David bowl) is a very attractive don footed bowl and is a tribute to the Jewish religion. The Star is interlaced and is beaded while a graceful bord of flowers, tendrils and bows edge the stippled center. The exterior is the Vintage pattern, and the colors a marigold, green and amethyst.

STAR SPRAY

Now and then this Imperial pattern is found in crystal or marigold Carnival glass; however, the smoke col is quite scarce, especially on the complete bride's basket. The bowl measures 7½" in diameter and has a beauti finish. The metal holder is nicely done, having a fine gold overspray and tiny rosettes with leaves on the handl

Star and Hobs

Star and File

Star Fish

Star Medallion

Star of David

Star of David and Bows

Star Spray

STARFLOWER

I certainly wish I could provide a manufacturer of this rare and beautiful pitcher (no tumblers known) bu can't. Known in both blue and marigold, the pitcher has turned up in two heights. The mold work is outstandi and the design flawless. Please notice how much the design resembles that of Millersburg's Little Star patte

STIPPLED ACORNS

The covered candy dish shown comes rather late in Carnival glass production but is still attractive enough be of interest. It stands 6½" tall on a 3½" footing. I've seen these in marigold, blue and amethyst, but don't kn who made them.

STIPPLED DIAMOND SWAG

This beautifully designed English compote seems to be the only shape known in this pattern, and while M Presznick reports green and blue ones, I've seen only the rich marigold shown. The compote is 5" tall a measures 5¾" across the top, with a 3½" base.

STIPPLED PETALS

While this Dugan pattern isn't uncommon in peach opalescent bowls, the pretty handled basket sets it w above the ordinary. Made from the 9" bowl, this basket is a large one and has been seen in amethyst as w as the beautiful peach shown.

STIPPLED RAMBLER ROSE

Most trails seem to lead to the Dugan company with this seldom-discussed pattern, but I have no proof th made it. The shape shown is the only one I've seen, and the colors are marigold and blue only.

STIPPLED RAYS (FENTON)

While Stippled Rays was used by both Northwood and Millersburg too, the Fenton version is perhaps the m commonly known and is available in bowls, bonbons, compotes, plates, creamers and sugars. The colors found a marigold, amethyst, green, blue and a rare red.

STIPPLED RAYS (IMPERIAL)

As you can readily see, the Imperial version of Stippled Rays is quite different than that of the Fent company in that the edges are evenly scalloped and the pieces are footed. Known only in a breakfast set to da I suspect other unreported shapes exist. The colors are marigold, smoke and helios green.

STIPPLED RAYS (NORTHWOOD)

Every Carnival-producing glass company had a stippled rays pattern, and it is probably the most comm motif in the entire field of iridized glass other than the grape. Northwood had several versions. The one sho is a variant. The most unusual aspects of the bowl are the reversed "N" and the exterior pattern which is a Gre Key and Scales. As you can see, it is a beautiful fiery amethyst and the bowl is dome footed. Stippled Rays w made in most colors and several shapes.

Stippled Acorns

Stippled Petals

Starflower

Stippled Diamond Swag

Stippled Rambler Rose

Stippled Rays (Fenton)

Stippled Rays (Imperial)

Stippled Rays (Northwood)

STIPPLED STRAWBERRY

While this pattern has been reported previously in a tumbler only, it obviously wasn't limited to that shape and it is a real pleasure to show this rare spittoon shape. It stands 3½" tall and measures 4½" across its widest part. The coloring is nothing spectacular but adequate. Along the spittoon's lip is a checkerboard pattern. The manufacturer is the Jenkins company.

STORK ABC PLATE

I'm certain many collectors will question my placing this pattern as an Imperial one. But on close examination this pattern and that of the Bellaire Souvenir bowl are almost identical in makeup. It is my guess both patterns were turned out late in the Imperial line and each are rather scarce. The only color is marigold.

STORK AND RUSHES

Stork and Rushes is another pattern whose shards were found in the Dugan diggings and since it has features that are typical of Northwood, I'm afraid it's difficult to state whether Dugan or Northwood made this pattern. At any rate, it is available in berry sets, punch sets, water sets, hats and mugs. I've seen only marigold, purple and blue with purple which is probably the hardest color to find.

STORK VASE

Much like the Swirl vase in concept, the Stork Vase was made by Jenkins and is usually found in a pale marigold glass. The Stork is on one side only, with stippling covering the other side. The Stork Vase stands 7½" tall.

STRAWBERRY (FENTON)

This little bonbon shape has been seen in all sorts of glass including crystal, custard, milk glass and Carnival glass where the colors range from marigold, cobalt, amethyst and green to the rare red amberina shown and good rich all red. The design qualities aren't all that good, but the iridescence is usually adequate and on the whole, the Fenton bonbon comes off nicely.

STRAWBERRY (MILLERSBURG)

Like its close relatives, Grape and Blackberry Wreaths, this beautiful pattern is the culmination of the design. Its detailing is much finer than either of the other patterns and the glass is exceptional. The coloring is a true grape purple, and the iridescence is a light even gold. Like other Millersburg patterns, the shapes vary, but the deep tri-cornered bowl with candy-ribbon edge is my favorite. A rare compote exists in the usual colors.

Stippled Strawberry

Stork ABC Plate

Stork and Rushes

Stork Vase

Strawberry (Fenton)

Strawbery (Millersburg)

193

STRAWBERRY (NORTHWOOD)

Available in only bowls of various sizes and plates in either flat or hand grip styles, the well-known Northwood pattern comes either plain or stippled. On the stippled version, there are three narrow rings around the outer edge; these are absent on the plain pieces. Colors are both vivid and pastel with the purple showing off the pattern to its best.

STRAWBERRY EPERGNE

The Strawberry Epergne is very much like the Fish Net one also made by Dugan; however the former is much rarer and has not been reported in peach opalescent yet. As you can see, the bowl exterior is plain and the base is domed.

STRAWBERRY SCROLL

Much like the rare Lily of the Valley in design, this rare Fenton water set is a real beauty that can stand along side of the best designs in Carnival glass. The shape, like that of the Bluebery set, is very artistic also. Colors reported are marigold and blue, but I suspect amethyst and green are possibilities.

STRAWBERRY INTAGLIO

I showed this pattern in *Millersburg, The Queen of Carnival Glass* as a questionable Millersburg pattern. Since then we've been able to trace it to the Northwood company through both a "goofus" bowl and a gilt decorated one, both bearing the famous trademark. The glass is thick, the design deeply impressed and the iridescnece only so-so. I've seen large and small bowls only.

STREAM OF HEARTS

The same heart shape employed by Fenton on the Heart and Vine, and Heart and Trees designs is found here on the scale fillers swirls that form a peacock's tail. Usually found on the compote shape, often with Persian medallion as an exterior companion, Stream of Hearts is available on a 10" footed bowl also. Colors are marigold, blue and white.

STRETCH

While we couldn't take the time to show all the shapes made in Fenton's vast stretch line, I wanted to show these matching pieces because they so typify the Fenton style and because they are so scarce. Please note the contrasting colored handles are not iridized.

Strawberry

Strawberry Epergne

Strawberry Scroll

Strawberry Intaglio

Stream of Hearts

Stretch

STRUTTING PEACOCK

The only shapes in this Westmoreland pattern are the creamer and sugar and the only colors I've heard abo are green or amethyst. The design is much like Shell and Jewel, also a Westmoreland product, but beware; the have been reproduced.

STUDS

Studs is found in a large tray, milk pitcher and footed juice tumbler. It is reported to be from Imperial, b I've always had a doubt about this. Marigold is the only color.

SUN-GOLD EPERGNE

Although I question the origin of this beautiful epergne, I don't for one minute doubt its desirability. The ba is of highly polished pierced brass while the bowl and lily are of an unusual pinkish-marigold Carnival glass. T epergne is 12" tall and the bowl has a 9¼" diameter. The glass is clear and mirror-like and has good ev iridescence.

SUNFLOWER

Sunflower must have been a very popular pattern in its day, for numerous examples have survived to t present time. It's quite easy to see why it was in demand. It's a pretty, well-designed Northwood pattern th holds iridescence beautifully. The bowl is footed and carries the very pleasing Meander pattern on the exteri It is also found, rarely, on a plate. The colors are marigold, green, amethyst and a rare teal blue.

SUNFLOWER AND DIAMOND

Long felt to be a Jenkins pattern, this well done vase is now known to be of Swedish origin and an examp in blue has been reported. The usual color is a good marigold. The pattern is all intaglio and very deep.

SUNFLOWER PINTRAY

Like its companion piece, the Seacoast Pintray, this Millersburg item joins the list of a select few. It is 5 long and 4½" wide and also rests on a collar base. The most unique feature is, of course, the open handle a the colors are marigold, purple, amethyst and a rich green.

SUNKEN HOLLYHOCK

Probably found more readily than the other "Gone With the Wind" lamps, Sunken Hollyhock can be found marigold (often with a caramel coloring) and a very rare red. The lamp stands an impressive 25" tall a certainly is a show stopper.

Studs

Strutting Peacock

Sun-Gold Epergne

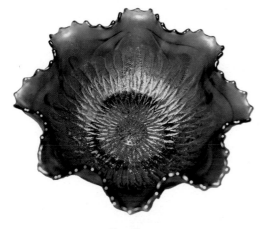

Sunflower

Sunflower and Diamond

Sunflower Pintray

Sunken Hollyhock

SUNRAY COMPOTE

These small compotes, found on marigold over milk glass, are very pretty in their own plain way. Fenton ma[de] many such items, and I strongly feel this compote came from that company also.

SUPERB DRAPE

The title of this beautiful piece of glass is certainly appropriate – it is superb! About 6½" tall and 7" diameter, this very rare vase is a true aqua with a rich even butterscotch iridization. The gently rolling top sho[ws] a mellow opalescence as does the base. All in all, this Northwood creation is a rare beauty that would grace a[ny] collection superbly!

SWAN, PASTEL

Of all the shards of patterns I catalogued from the Dugan dump site, I'm sure this pattern surprised me mo[re] than any other. For years I considered these small novelties either Fenton or Westmoreland products with [my] vote leaning toward the latter. However, we now can say with some positiveness that they are Dugan. Found [in] pink, ice blue, ice green, marigold, peach and purple. The darker colors aand peach are the scarcest and dema[nd] greater prices. A variant by Fenton is known, however.

SWEETHEART

Besides the rare and beautiful covered cookie jar shown, this Cambridge pattern is known in a very ra[re] tumbler shape in marigold. The cookie jar has been seen in marigold, green and amethyst and as you can se[e] the mold work and finish are superior.

SWIRL (IMPERIAL)

This 7" vase is shown in old Imperial catalogs, and I've heard of it in marigold, smoke, green and white. T[he] design is nothing outstanding, but the useful shape and the iridescence are adequate.

SWIRL (NORTHWOOD)

I've seen this pattern on a beautiful tankard water set and the mug shape in marigold. The tumbler is sho[wn] in the Owens book in green and is known in amethyst. While the tumblers are often marked, the mug isn't. No[w] and then, the tankard pitcher is found with enameled flowers added. Naturally the pitcher is scarce and the mu[g] is considered rare.

Sunray Compote

Superb Drape

Swan Pastel

Sweetheart

Swirl (Imperial)

Swirl (Northwood)

SWIRLED FLUTE VASE

This little Fenton vase is a real charmer for such simplistic design. The wide panel is quite pronounced at t[...] base and the color and iridescence are quite good. These little vases average about 9" in height and are fou[...] in red, marigold, green, amethyst, blue and white.

SWIRL CANDLESTICK

This very attractive candlestick is the same pattern as the rare mug I showed in previous books and w[...] apparently a product of either Northwood or Dugan. While other colors may exist, I haven't heard of any. T[...] luster is top notch and the glass, quality all the way.

SWIRL HOBNAIL

This is a little Millersburg jewel in either the rose bowl or the lady's spittoon, and once you own either pie[...] wild horses couldn't drag it away. The glass is simply sparkling and the iridescence outstanding. The spitto[...] has an irregular scalloped edge opening and both shapes have a many-rayed base. The usual colors prevail, b[...] the green is extremely difficult to locate. Swirl Hobnail can also be found in a vase shape.

SYDNEY

The rare Sydney tumbler shown is one of a handful known and is from the Fostoria Glass Company. The on[...] color reported is marigold that tends toward a honey-amber tone. The four large ellipsed crosses around t[...] tumbler are the main motif and this design is repeated on the tumbler's base.

TAFFETA LUSTRE CANDLESTICKS

These very rare Fostoria candlesticks were manufactured in 1916 or 1917 (according to an old Fostoria catalo[...] in colors of amber, blue, green, crystal or orchid. They were part of a "flower set" which included a centerpie[...] bowl 11" in diameter. The candlesticks themselves came in 2", 4", 6", 9", and 12" sizes, and as you can see, the[...] shown still have the original paper labels on the bottom. When held to the light, the ultra-violet color is fantast[...] and the iridescence is heavy and rich. Fostoria made very small amounts of iridized glass and certainly the[...] examples of their Taffeta Lustre line are quite rare.

TEN MUMS

Found on beautiful water sets, large impressive bowls, and rare plates, Fenton's Ten Mums is a very realist[...] pattern. The mold work is unusually fine, especially on the bowl shape. Colors are marigold, cobalt blue, gree[...] peach opalescent and white but no all shapes are found in all colors.

THIN RIB (NORTHWOOD)

Here is an exterior pattern every one must know for besides the famous basketweave, it is found on some[...] Northwood's best known bowl patterns, including Good Luck and Strawberry. It is simple enough but held to t[...] light, it adds greatly to the design.

THIN RIB AND DRAPE

Much like the other Thin Rib vases, this one has an interior drape pattern, adding to the interest. It can [...] found in several colors including marigold, green and amethyst. I suspect Fenton is the maker, but can't be sur[...]

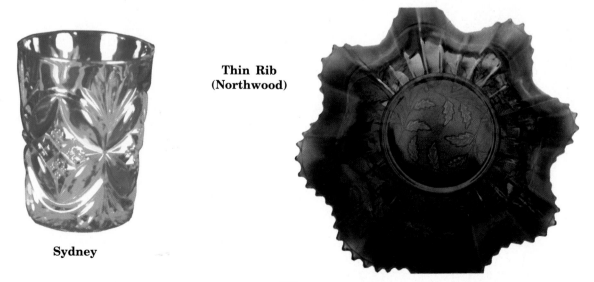

**Thin Rib
(Northwood)**

Sydney

Swirled Flute Vase

Swirl Candlestick

Swirl Hobnail

Taffeta Lustre Candlesticks

Ten Mums

Thin Rib and Drape

THISTLE

Fairly typical of many Fenton bowl patterns, Thistle is artistically true with the thistles and leaves realist and graceful. Found on bowls, plates and a rare compote, the colors seen are marigold, green, blue a occasionally amethyst. Now and then a bowl appears with advertising on the base.

THISTLE AND THORN

British in origin, this nicely designed pattern is found on a variety of shapes, all footed, including bow sugars, creamers, plates and nut bowls. Colors are usually marigold but I've had blue reported.

THISTLE BANANA BOAT

This beautiful Fenton banana boat is another of the underrated patterns in Carnival glass. Massive in conce bold in design with the thistle on the interior and cattail and waterlily outside, the four-footed banana boat usually found on marigold, green or cobalt blue. The iridescence is quite heavy, usually with much gold.

THISTLE VASE

What a pretty little vase this is. Standing 6" tall with a soft amber shading to the glass, the Thistle vase a well-designed bit of color for vase lovers. The maker is unknown.

THREE FRUITS

So very close in design to the Northwood Fruits and Flowers pattern shown earlier in this book, the t substantial differences are the absence of the small flowers and the addition of an extra cluster of cherries to th pattern. Found in bowls of all sizes, including flat and footed ones, and average size plates. Northwood's Thr Fruits is available in all vivid colors as well as pastels.

THREE FRUIT MEDALLION

If you will look closely you'll see the differences between this pattern and the regular Three Fruits desig including the addition of flowers and leaf tendrils. The large bowl is footed, has a reverse pattern of Meand and the one shown is on black amethyst glass and is a real show-stopper.

THREE FRUITS VARIANT

If you'll compare this beautiful 9" plate with the Northwood version, you'll see a good deal of similarities a some obvious differences. While this version has been credited to Fenton, I strongly suspect it was a Dug product.

THREE-IN-ONE

Originally called "Number One" in the crystal line, this well-known Imperial bowl pattern (plates also exis is always easy to spot because of the two rows of near-cut diamond designs, separated by a center area of flute I've seen bowls in 4½", 7½", 8¾" sizes, in colors of deep marigold, green, purple and smoke. The iridescence usually quite good and the glass heavy, clear and sparkling.

THREE-IN-ONE VARIANT

This beautifully made toothpick holder has been iridized on only the center sections and is exactly like t Three-In-One pattern except for the file work within the top and bottom row of diamonds. Beyond any questi it is an old one.

Three Fruits Medallion

Three-In-One Variant

Thistle

Thistle and Thorn

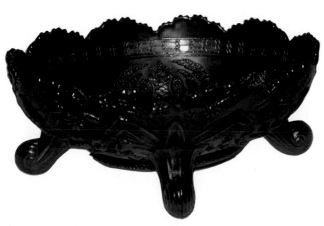

Thistle Banana Boat

Thistle Vase

Three Fruits Variant

Three-In-One

Three Fruits

THREE MONKEYS BOTTLE

I'm not sure how many of these are around but not many I'd guess. The one shown is the reported exam in all the previous pattern books, so it may have only a few brothers. The Three Monkeys bottle is of clear gl with good iridescence, stands 8" tall and bears the words, "Patent Pending – 8 oz." on the base. A rare find bottle lovers.

THREE ROW VASE

I was very impressed when I saw this beautiful Imperial rarity. The color was truly fabulous and iridescence fine enough to rival the Farmyard bowl. I have heard of only this one piece in beautiful purple, others may exist. It is 8" tall and 4½" wide, and the mold work is quite deep and sharp. There is also a varia called Two Row vase which is quite similar.

THUMBPRINT AND OVAL

Standing only 5½" tall, this well-molded Imperial pattern is known in marigold and purple. Either color scarce and Thumbprint and Oval is much sought.

THUNDERBIRD (SHRIKE)

Called Thunderbird in America, the bird shown in actually a shrike. The flora is, of course, wattle. T Australian pattern can be found in both large and small bowls in marigold or purple.

TIGER LILY

It really is a shame this well-done Imperial pattern was used only on the water set. It would have been qu effective on any numer of other shapes, including table sets and punch sets. The mold work is some of Imperia best, all intaglio, and very sharp and precise. Tiger Lily is found in marigold, green and rarely purple. It has be reproduced in pastels, so beware!

TINY COVERED HEN (LITTLE)

The Imperial novelty item shown is a really well-conceived miniature that until recently had not be reproduced; however, I saw one of these in a gift shop in blue glass (not Carnival) not long ago. At any rate, t one shown is old and rare. The coloring is a very nice clambroth, richly iridized with a fiery finish. It measu 3½" long and 3¼" tall.

TINY HOBNAIL LAMP

In all my years in collecting, I've seen just two of these miniature oil lamps even though they do come la in Carnival glass history. The base is 3" high and has a diameter of 2¼". The shade, which matches, is cle I haven't a notion as to the maker, but they are scarce and cute as can be.

TOBACCO LEAF (U.S. GLASS)

Here is another souvenir champagne, but the one shown is a real rarity. Please note that the finish is go not the usual clear iridized! As the owner says, this must have been for someone very special at the conventi

TOMAHAWK

This lovely miniature (7¼" long and 2" wide) is a real "show stopper." Now known to have been a product the Cambridge Glass Company, the rare Tomahawk is a very deep cobalt blue with heavy iridescence. It has be seen in pieces other than Carnival glass and has been reproduced in aqua and vaseline glass, but as far as know, no iridized reproductions were made.

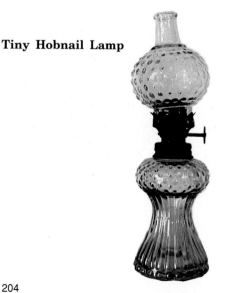

Tiny Hobnail Lamp

Tobacco Leaf (U.S. Glass)

Three Monkeys Bottle

Three Row Vase

Thumbprint and Oval

Tiger Lily

Thunderbird (Shrike)

Tiny Covered Hen

Tomahawk

TORNADO

Perhaps because it is one of the most unusual vases in all of Carnival glass, Northwood's Tornado is alwa a favorite with collectors. Available in both plain and ribbed, the size may vary considerably, and I've seen a mi version in marigold. The colors are marigold, green, purple, white and ice blue. The Northwood trademark found on the inside of the vase.

TORNADO VARIANT

This extremely rare variant is one inch taller that the regular Tornado vase. The base closely resembles th of Northwood's Corn vase and the top is tightly crimped. The only color reported is a deep rich marigold, iridiz both inside and out.

TOWN PUMP

Certainly there isn't one collector of Carnival glass who isn't familiar with this very famous pattern. The To Pump is almost 7" tall and is mostly on purple, although marigold and green are found in limited quanities. T design is simple but very pleasing – ivy twining over a stippled background with a crude tree bark spout a handle. No Carnival glass collection would be complete without this pattern and certainly it deserves a promin place in any Northwood collection.

TRACERY

This pattern comes as quite a surprise to many Millersburg fans because of its delicate patterning, b Millersburg it certainly is. It is found on bonbons of rather deep, oval shape and is usually seen in green amethyst. It measures 7½" long and 5½" wide. The base is 3" in diameter and has a many-rayed center. T exterior is plain and has two mold marks.

TREE BARK

As you can see, this is a simplistic pattern, turned out as a giveaway or inexpensive item. The only shap I've heard of are water sets with either open or lidded pitchers and the only color is marigold, usually of a de rich hue. While it certainly does not show the artistry the Imperial company was known for, Tree Bark does ful its purpose of being an available bit of color for the average housewife's table.

TREE BARK VARIANT

If you look closely, you can see this design is slightly different than the regular Tree Bark pattern, havi grooves that are straight rather than random spaced. I've seen this on this small tumbler as well as an orna candle holder with a fancy brass cherub and a marble base.

TREE OF LIFE

There has always been a great deal of confusion surrounding this pattern and Crackle and Soda Gold patter because they are so similar, but there is really no need for the confusion. As you can see, Tree of Life has stippling whatsoever, thus the filler area is plain. The pitcher shown is a scarce shape in this pattern, and you can see, the iridescence is thin and light. The shapes known are water sets, plates and perfume bottle

TREE TRUNK

Simple but effective is the bark-like pattern for this popular North-wood vase. Apparently it was a well-liked pattern when it was being manufactured for many examples exist in sizes from 8" to tall 20" ones. The colors are marigold, blue, purple, green, white and aqua opalescent.

TRIANDS (AND TOWERS VASE)

Triands is a product of Sowerby and is found in a covered butter dish, creamer, sugar, celery and spooner (both shown). The only color reported is a good rich marigold. The other small hat shape shown is called Towers and is a vase pattern also from Sowerby.

Triands (and Towers Vase)

Tornado

Tornado Variant

Town Pump

Tracery

Tree Bark

Tree of Life

Treebark Variant

Tree Trunk

TRIPLETS (DUGAN)

This common Dugan patter is probably seen as much as any by the beginning collector and must have be produced by the carloads. Colors are marigold, amethyst, green, peach opalescent and white but not all are g and some of the peach is downright awful.

TROPICANA VASE

The maker of this intriguing vase is unknown, but I suspect it may be English. It stands 9" tall and ha very rich marigold coloring.

TROUT AND FLY

Of course, this is a companion piece to the Big Fish pattern and except for the added fly and a few chang in the water lilies and blossoms, is much the same. The bowls usually measure about 9¼" in diameter and ha a wide panel exterior. The detail is good and the coloring excellent. The shape may vary and even square bow and a rare plate shape have been found in this Millersburg pattern.

TULIP (MILLERSBURG)

This beautiful and rare compote is exactly like the Flowering Vine one shown elsewhere except it has interior design. It stands about 9" tall and the bowl measures roughly 6" across. The only colors I've seen a marigold and amethyst but green was more than likely made. In addition to this size, a 7" variant is known amethyst.

TULIP AND CANE (IMPERIAL)

Seldom found, this old Imperial pattern is seen in iridized glass only on two sizes of wine goblets, a cla goblet and a rare 8 oz. goblet, in marigold and smoke.

TULIP AND CANE (MILLERSBURG)

Only a few of these attractive Millersburg vases seen to show up from time to time, mostly in amethyst, b once in a while in green or the quite rare marigold. I've seen them as tall as 11" but most are about 7" and le stretched out. They are always quickly snatched up and are a popular vase pattern.

TWINS

Like Imperial's Fashion pattern, Twins is another geometric design manufactured in large quantities in ber sets, a rare bride's basket, and the familiar fruit bowl and stand. Pieces have been found in a good rich marig and a very beautiful smoke and the berry set has been found in green, as shown. Needless to say, a purple fr bowl and stand would be a treasure!

TWO FLOWERS

Once again the scale filler, water lilies, and blossoms are used by Fenton to form an artistic presentatic Known in footed and flat bowls of various sizes and a scarce bowl shape, Two Flowers is seen in blue, marigo amethyst, red, green and occasionally white.

TWO FRUITS

Once this Fenton pattern was felt to be an Imperial product but we know from old advertisements that it came from Fenton. The background is Fenton's Stippled Rays and, as you can see, there are four sections to the divided bonbon. Each section contains one fruit – either a pear or an apple. Colors I've heard about are marigold, blue and amethyst but certainly green may exist.

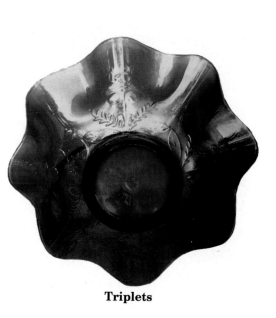

Triplets

Tropicana Vase

Trout and Fly

Tulip and Cane

Tulip Scroll

Tulip (Millersburg)

Twins

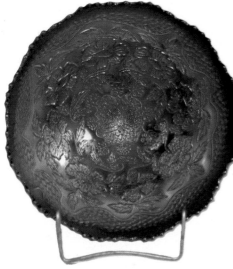

Two Flowers

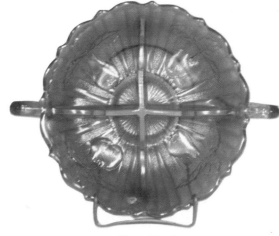

Two Fruits

TWO FRUITS

Here is a Northwood pattern so unique and so very rare, it is not listed in the two major pattern guide. Shown is the spooner in a rich cobalt blue. It last sold in the 1977 Wishard auction. I know of one other pie in this pattern, a lidless sugar. Shown in the cherry side; the other pattern on the opposite side shows t peaches. The spooner measures 6¼" x 4½" and bears the Northwood trademark.

UNSHOD

The owner says this pitcher came with tumblers in the Unshod pattern so we've listed the pitcher as th pattern. If anyone knows differently, I'm sure I'll hear about it. The styling look very much like the Millersbu Cherry, but that's only an observation.

URN VASE

This unusually shaped vase has been seen in white as well as marigold Carnival glass and is a product of t Imperial company. It stands 9" tall and has a finely grained stippling over the surface.

VENETIAN

While credited to the Cambridge company, there is evidence that this pattern was iridized at the Millersbu factory. It is shown in old Cambridge ads, however, and the vase was used as a lamp base. Colors are green a a scarce marigold. The base design is quite similar to the Hobstar and Feather rose bowl. A rare butter dish shown in marigold.

VICTORIAN

This impressively large bowl is another pattern from the Dugan factory in Indiana, Pennsylvania and is bo rich and desirable. Mostly found in a rich purple, one example in a rare peach opalescent has turned up and rat with the best of bowls.

VINEYARD

Here is another pattern whose shards I catalogued from the Dugan diggings, but actually it didn't come too much of a surprise. The same design was made in 1905 in crystal, opaque glass, and opalescent ware und the name, Grape and Leaf. Known in iridescent glass, in a water set only, the colors are marigold, purple a rarely peach opalescent. Often the tumblers are poorly formed and the glass tends toward being bubbled.

VINING LEAF

There are two variants of the English pattern – one with small berries along portions of the stylized leave The examples shown, a rather small bud vase and an ample lady's spittoon, give us the pattern nicely. Plea note the frosted effect around the leaves.

VINTAGE (FENTON)

Vintage was the Fenton company's primary grape pattern, and as such can be found in a wide range of shape including bowls, plates, rose bowls, compotes, punch sets, ferneries and a one-lily epergne. Shown is the Vinta fernery in a rare red. Other colors known are marigold, blue, green, amethyst, amber, pastel marigold a amberina.

VINTAGE (MILLERSBURG)

What makes Vintage unique is the hobnail exterior (the hobnail and the honeycomb seem to have be patterns used as standard fillers by Millersburg). It has a many-rayed star in the base and is usually seen the medium size, shallow bowls. The Millersburg Vintage pattern is a scarce one and is certainly a credit to a collection. Colors are green, marigold, amethyst and blue. And both 9" and 5" bowls are known.

VINTAGE BANDED

Often seen in the mug shape, this Dugan pattern is rather scarce in the footed pitcher shape, and the tumbl is rare and much sought. Please note the grapes are almost identical to those found on the Golden Grape patter and the banding is quite similar to that found in the Dugan pattern called Apple Blossoms.

Vintage (Millersburg)

Vintage (Fenton)

Two Fruits

Urn Vase

Venetian

Victorian

Unshod

Vineyard

Vining Leaf

Vintage Banded

VINTAGE LEAF (FENTON)

Over the years, I've never been able to accept this design with the center leaf as the standard Fenton Vinta[ge] pattern so I've given in to my conscience and given it a name all its own. It is found in berry sets in marigo[ld] green, blue and amethyst and is Fenton, not Millersburg, despite the radium finish.

VIOLET BASKETS

Here are two of the dainty little baskets designed to hold small bouquets like violets. One has its own gla[ss] handle, while the other fits into a handled sleeve. The File pattern in the glass is much like that on some Sto[rk] and Rushes pieces, and I suspect both of these baskets come from the Dugan company.

VIRGINIA BLACKBERRY

Known to be a variant of a U.S. Glass pattern called Virginia or Galloway, this small pitcher has turned [up] on a child's water set in crystal without the berries. The example shown in blue Carnival is the only report[ed] one that is iridized, making it a very rare item indeed.

VOTIVE LIGHT

Also bearing the strange "M-inside-a-C" mark like Oklahoma, this very hard-to-find novelty has a beautif[ul] marigold finish. It stands 4¼" tall and is a candle vase. The side shown is the Bleeding Heart of Jesus. On th[e] opposite side is a raised cross measuring nearly 2" tall.

WAFFLE BLOCK

This Imperial pattern was originally made in many shapes. The pitcher is hard to come by and, of course, th[e] tumblers are rather are, but other shapes, including a handled basket, rose bowl, punch set, vases, bowls, parfa[it] glass and shakers are known. Colors are marigold, clambroth, aqua and amethyst.

WATER LILY

Found on bonbons and flat and footed berry sets, Water Lily is similar to Lotus and Grape, Pond Lily ar[d] Thistle and Lotus. They each give a feeling of a small lily pond. The usual lotus-like flowers are present alor[g] with water-lily blossoms, leaves and small vine-like flora. Colors known are marigold, blue, green, amethys[t], amber, white and red.

WATER LILY AND CATTAILS (FENTON)

While both Fenton and Northwood had versions of this pattern, the Fenton one is best known and is availab[le] in more shapes, including water sets, table sets, berry sets and a rare spittoon whimsey. Colors are marigold ar[d] amethyst, but certainly blue or green may exist but have not been reported.

Vintage Leaf

Virginia Blackberry

Violet Baskets

Votive Light

Waffle Block

Water Lily

Water Lily and Cattails

WATER LILY AND CATTAILS (NORTHWOOD)

Here is a pattern used in different degrees by both the Northwood and Fenton companies, and at times it rather hard to tell who made what. The obvious Fenton rendition is the exterior pattern used on the Thist banana boat, but it is believed they made other shapes in Water Lily and Cattails, including a whimsey toothpic This is the Northwood water set.

WATER LILY AND DRAGONFLY

This large (10½") shallow bowl and matching frog is called a flower set and was used much like an epergn A large flannel flower is under the frog. In Australia these are also called "float bowls."

WEBBED CLEMATIS

I apologize for the weak photo of this hugh vase (12½" tall) but it was the only one available. I suspect th vase is Scandanavian and it reminds me of several others from that area. The color is light marigold.

WHEAT

Make no mistake about it, this is a very rare Northwood pattern, known in only a covered sweetmeat (tw known) and a single covered bowl in colors of green and amethyst. Surfacing in Carnival glass circles only in th last 10 years, these rarities have caused much excitement and brought astonishing prices where they've bee shown. The sweetmeat is the same shape as the Northwood Grape pattern. What a pity we don't have more this truly important pattern.

WHIRLING LEAVES

What a simple but effective pattern for large bowls! Four flowing leaves and four star-like blossoms – wha could be simpler? Please note that the flowers are very much like those found on the Little Stars bowls. Th pleasing pattern is not especially difficult to find and does not sell for nearly as much as most Millersbu patterns but it is worth owning.

WHIRLING STAR

While this Imperial pattern had never brought many raves, it is a nice, well-planned design found on punc sets, bowls and an attractive compote. Colors are mostly marigold, but an occasional piece in green is found.

WHITE OAK

After years of doubt about the maker of this rare tumbler (no pitcher known), I'm half convinced it was Dugan product. The bark-like background and the general shape are much like the Vineyard tumbler. Of cours I could be wrong.

WICKERWORK (ENGLISH)

Sometimes only the bowl without the base of these is found, but here is the complete set. The bowl measure 8½" in diameter. The only color I've seen is the marigold. Apparently, this was a cookie or sandwich tray.

WIDE PANEL (IMPERIAL)

Like the other major Carnival glass companies, Imperial produced a Wide Panel pattern. These are found primarily in goblets of two sizes and a covered candy jar. Colors known are marigold, ice blue, ice green, vaseline white, pink and red. Most have a stretch effect around the edge.

Webbed Clematis

Wickerwork (English)

214

Waterlily and Cattails

Water Lily and Dragonfly

Whirling Star

Whirling Leaves

Wheat

White Oak

Wide Panel

WIDE PANEL EPERGNE

This beauty is the most stately epergne in Carnival glass and is a product of the Northwood factory. Col_ are white, marigold, green, amethyst and the ultra-rare aqua opalescent one shown.

WIDE PANEL VARIANT

Rather plain for a Northwood product, this water set relies on symmetry and grace for its charm. It can _ found in marigold and white as well as enameled in both colors and a beautiful persian blue.

WIDE PANELED CHERRY

Unlisted until now, this very interesting pattern is similar to the Stippled Strawberry spittoon I showed _ *Rarities In Carnival Glass* some years ago. The shape is a very small pitcher, possibly used for syrup or cre_ and to date I've heard of no matching pieces.

WIDE RIB VASE

While several glass companies produced these wide rib vases, the one shown came from the Dugan facto_ Colors I've seen are marigold, green, blue, amethyst, smoke, white, amber, peach opalescent and aqua opalesce_

WILD BLACKBERRY

Very much in design like the other Fenton Blackberry pattterns, this is strictly a bowl pattern easily identif_ by the four center leaves and the wheat-like fronts around the outer edges. Colors are marigold, blue, green a_ amethyst. The exterior has the wide panel pattern.

WILD FLOWER (MILLERSBURG)

This deeply cupped, clover-based compote is just over 6" tall and is 4½" wide. The very graceful pattern of fo_ large leaves, four small leaves, and eight blossoms is quite effective. The iridescence is found only on the insi_ of the compote and has the usual radium finish. This is not an easy compote to find, and anyone having o_ should be proud. This Millersburg pattern is found in marigold, green, amethyst and a rare vaseline.

WILD FLOWER (NORTHWOOD)

The only place I've ever seen this pattern is on the outside of Northwood's Blossomtime compotes and tha_ a real shame. It is a beautiful pattern and certainly deserves better.

WILD GRAPE

I recently encountered this pattern for the first time and at first glance thought it to be a Palm Beach bow_ but as you can clearly see, it isn't. The color is weak, but the design is good and the dome base looks like ma_ Dugan bowls.

WILD LOGANBERRY (DEWBERRY)

Since I first showed this Westmoreland pattern in the pitcher and announced the existence of wine a_ goblets, other shapes have turned up including a covered compote, creamer, sugar and an open compote. _ except the marigold wine have been iridized milk glass, but other colors may exist.

WILD ROSE

This well-known Northwood pattern is used on the exterior of two types of bowls. The first is an average f_ bowl, often with no interior pattern. The second, however, is an ususual footed bowl, rather small, with an ed_ of open work in fan-like figures. Certainly, the intricacy of this open work took much care and skill. The col_ are marigold, amethyst, green and occasionally pastels.

**Wide
Paneled
Cherry**

Wild Grape

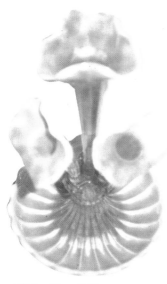

Wide Panel Epergne

Wide Rib Vase

Wild Blackberry

Wild Flower

Wildflower (Northwood)

Wild Loganberry Pitcher

Wild Rose

Wide Panel Variant

WILD ROSE LAMP

This very scarce lamp is found in three sizes and three colors, amethyst, green and marigold. The font is clear glass but the remainder of the lamp is well-iridized. The Wild Rose pattern is well raised and meanderi around the base of the lamp is a pattern of sunken dots. Occasionally one of these lamps is found with thr medallions on the underside. These medallions contain portraits of three ladies believed to be wives of the thr major stockholders of the Millersburg company. One of these lamps would be a credit to any glass collection a certainly deserves top billing in a collection of Millersburg glass.

WILD ROSE SYRUP

Seen only in marigold of rich color and adequate lustering, the Wild Rose Syrup is very attractive. The mak is thus far unknown, but the mold work is quite similar to that of many of Harry Northwood's produc Measuring 6½" tall, the syrup holds 12 fluid ounces and has a metal top.

WILD ROSE WREATH (MINIATURE INTAGLIO)

The Butler Brothers ad showing this rare and desirable miniature call it a "stemmed almond" so we know was intended as a nut cup and was made by U.S. Glass. It stands 2½" tall and is usually about 3" across t top. The design is intaglio and the only color reported is marigold.

WILD STRAWBERRY

One might think of this as the "grown-up" version of the regular Northwood Strawberry since it is much t same. The primary difference is the addition of the blossoms to the patterning and, of course, the size of the bo itself, which is about 1½" larger in diameter.

WINDFLOWER

Not one, but two shards of this pattern were excavated at the Dugan site, including a marigold bit and sizeable chunk of an ice blue nappy with a portion of the handle attached. To the best of my knowledge, th pattern has not been reported previously in pastels. Windflower is an uncomplicated, well-balanced design wi the background stippled and a geometric bordering device. The exterior is plain. Windflower is found on bow plates and handled nappies. The colors are marigold, cobalt, ice blue and ice green.

WINDMILL

Again, here is one of the best-selling Imperial patterns, reproduced in the 1960's. It is still rather popular wi collectors of old Carnival glass, especially in the darker colors. Originally, Windmill was made in a wide ran of shapes, including berry sets, water sets, fruit bowls, pickle dishes, dresser trays and milk pitchers. The colo were marigold, green, purple, clambroth and smoke.

WINE AND ROSES

Please note the similarities between this scarce Fenton cider set, the well-known Lotus and Grape patte shown elsewhere in this book, and the very rare water set called Frolicking Bears. Wine and Roses can be fou in marigold in both pitcher and wine, and the wine is known in blue and the rare aqua.

WISHBONE

I personally feel that this Northwood pattern is one of the most graceful in Carnival glass. The lines just flow, covering much of the allowed space. Found on very scarce water sets, flat and footed bowls, plates and a very stately one-lily epergne, Wishbone is usually accompanied by the bas- ketweave pattern on the exterior. My favorite color is the ice blue, but Wishbone is available in marigold, green, purple, blue, white and ice green. There is also a variant to this pattern.

Wild Rose Lamp

**Wild Rose Wreath
(Miniature Intaglio)**

218

Wild Rose Syrup

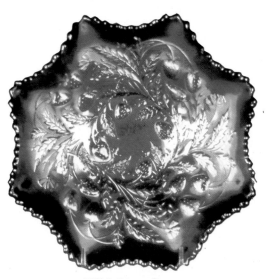

Wild Strawberry

Windflower

Windmill

Wine and Roses

Wishbone

WISHBONE AND SPADES

I've always had misgivings about attributing this pattern, and I'm listing it here as a questionable Dugan c As you can see, the design is a good one, well balanced and artistically sound. The shapes are berry sets a large and small plates and colors include peach opalescent, purple and green.

WISHBONE FLOWER ARRANGER

Similar in many ways to the Windsor Flower Arranger, for years I confused the two until one collector v kind enough to point out the differences. The Wishbone has an open top and none of the points touch while t isn't true with the Windsor.

WISTERIA

What a beautiful pattern this is! Certainly a first cousin to the Grape Arbor pattern in iridescent glass as v as the Lattice and Cherry pattern in crystal, Wisteria is unfortunately found only on water sets. While o tumblers in ice green have surfaced, both pitchers and tumblers are known in white and a really outstanding blue. What a shame no vivid colors are available in this Northwood masterpiece.

WOODEN SHOE

It seems as if good things come in bunches, and that's how it's been with miniatures in this book. Howev one could never tire of seeing them and they are so rare, so here is another. The Wooden Shoe is 4½" long a 3" high. Its coloring is a light watery amber, and the glass is heavy.

WOODLANDS VASE

Once again I show a rare vase that is very impressive even though it measures only 5½" tall. The color is a rich marigold with heavy luster and the pattern is all smooth and raised. The design has a simple, w balanced attractiveness that raises it above the everyday. And, of course, not many are known.

WOODPECKER VASE

The Woodpecker vase is 8¼" long and measures 1⅝" across the top. The only color I've run into is marig usually of good color. Perhaps this is strictly a Dugan pattern, and if so, it would explain the lack of other co or shapes. These vases were usually hung in dooways or used as auto vases and were quite popular in their d

WREATHED CHERRY

For many years, collectors have been puzzled by certain pieces of this pattern turning up with the Diamo Glass Company trademark, notably the covered butter. Now, since the Dugan shards have been catalogued becomes apparent this was a Dugan pattern. Known in berry sets consisting of oval bowls, table sets, water s and a scarce toothpick holder, the colors are marigold, fiery amethyst, purple and white (often with gilt). T toothpick has been widely reproduced in cobalt, white and marigold, so beware! The only color in the old o is amethyst.

WREATH OF ROSES (FENTON)

Known in bonbons with or without stems, compotes, bowls and the punch sets shown. Fenton's Wreath Roses is a beautifully executed pattern. Two interior patterns are known. Colors are marigold, green, bl amethyst, peach opalescent and white. The luster is very fine and the mold work excellent.

Wishbone Flower Arranger

220

Wishbone and Spades

Wisteria

Wooden Shoe

Woodlands Vase

Woodpecker Vase

Wreathed Cherry

Wreath of Roses

WREATH OF ROSES ROSE BOWL (DUGAN)

I've never understood while this little rose bowl has been grouped with the Fenton Wreath of Roses as (and the same pattern when it is obviously so different. So when a large hunk of Wreath of Roses rose bowl v found among the Dugan shards, the explanation became obvious. They simply are not the same pattern! Fou in marigold and amethyst, the Wreath of Roses rose bowl has superior mold work and adequate finish.

WREATH OF ROSES VARIANT (DUGAN)

While Northwood, Fenton and Dugan all made similar patterns with this name, the Dugan is least known valued. It hasn't all that much going for it since the design is weak and doesn't fill the space to advantage. Col are amethyst, blue and marigold – at least the ones I've seen, but white, green and peach opal may exist.

ZIG-ZAG (FENTON)

I suspect the same mold shape was used for this nicely done Fenton water set and the Fluffy Peacock patte The Zig-Zag was brought along in the enameled Carnival period and can be found in marigold, blue, gre amethyst, ice green and white. The floral work may vary slightly from piece to piece.

ZIG-ZAG (MILLERSBURG)

This lovely Millersburg bowl pattern is an improved version of a stippled ray theme but with a twist, result in a beautiful sunburst effect. This, coupled with a curious star and fan design on the exterior base, create unique and intriguing pattern. Colors are green, marigold and amethyst.

ZIPPERED HEART

Recent evidence that this is an Imperial pattern is a catalog from the company illustrating the Zippered He pattern in crystal in many shapes, including table sets, vases, compotes, rose bowls, a milk pitcher, punch s and water sets. However, in iridized glass, the shapes I've seen are a berry set and the famous Queen's va While a beautiful purple is most encountered, marigold is known. Shown is a rare 5" vase.

ZIPPER LOOP

While the Zipper Loop lamp is the most frequently encountered of all the kerosene lamps, it is a scarce it in itself. Known in four sizes, including a small hand lamp, the Zipper Loop is found in a good rich marigold co as well as a sparkling smoke finish. It has been reproduced in the former in the large size so be cautious wh buying.

ZIPPER STITCH

This attractive piece shown is the tray for a lovely cordial set consisting of a decanter, six stemmed cordi and the tray. These were made in Czechoslovakia and are known only in marigold.

ZIPPER VARIANT

I suspect this nicely done covered sugar is of British origin, but can't be sure. The color is quite good as the mold work. Note the interesting finial on the lid.

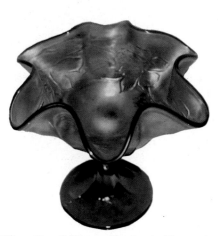

Wreath of Roses Variant (Dugan)

Wreath of Roses
Rose Bowl

Zig-Zag (Fenton)

Zig-Zag (Millersburg)

Zippered Heart

Zipper Stitch

Zipper Loop

Zipper Variant

The Standard
Carnival Glass
Price Guide

The current values in this book should be used only as a guide. They are not intended to set prices, which vary from one section of the country to another. Auction prices as well as dealer prices vary greatly and are affected by condition as well as demand. Neither the Author nor the Publisher assumes responsibility for any losses that might be incurred as a result of consulting this guide.

Color Code

Note: All purple items are found in the Amethyst column.

M--Marigold
A--Amethyst
P--Purple
B--Blue
G--Green
PO--Peach Opalescent
AO--Aqua Opalescent
BO--Blue Opalescent
IG--Ice Green
IB--Ice Blue
SM--Smoke
CL--Clear
CM--Clambroth
V--Vaseline
R--Red
Pas--Pastel
AQ--Aqua
AM--Amber
IM--Iridized Moonstone
HO--Horehound

PL--Pearl Opalescent
AB--Amberina
RA--Reverse Amberina
LV--Lavender
IC--Iridized Custard
VO--Vaseline Opalescent
PM--Pastel Marigold
W--White
TL--Teal
PeB--Persian Blue
CeB--Celeste Blue
AP--Apricot
BA--Black Amethyst
LG--Lime Green
RG--Russet (olive)
PK--Pink
MO--Milk Glass Opalescent
LO--Lime Opalescent
RS--Red Slag
WS--Wisteria

*Denotes a speculative price for items that are rare
and have not been on the market for some time.

	M	A	G	B	PO	AO	Pas	Red
ACANTHUS (IMPERIAL)								
Bowl, 8"-9½"	65	90	110	95			85SM	
Plate, 10"	275						290SM	
ACORN (FENTON)								
Bowl, 7"-8½"	40	65	80	56	290	450	390IB	620
Plate, 9", Scarce		490		480				
ACORN (MILLERSBURG)								
Compote, Rare	2,800	1,500	1,500				2500V	
ACORN BURRS (NORTHWOOD)								
Bowl, Flat, 10"	90	130	160				420	
Bowl, Flat, 5"	32	50	55				170	
Punch Bowl and Base	450	680	750	1,800		20,000	4,000	
Punch Cup	28	35	40	50		350	60	
Covered Butter	225	325					550	
Covered Sugar	195	230					375	
Creamer or Spooner	190	220					340	
Water Pitcher	465	690	850					
Tumbler	60	85	85					
Whimsey Vase, Rare	2,000	2,250						
ACORN AND FILE								
Ftd Compote, Rare							1000V*	
A DOZEN ROSES (IMPERIAL)								
Bowl, Footed, 8½"-10", Rare	525	575	595					
ADVERTISING ASHTRAY								
Various Designs	45+							
AFRICAN SHIELD (ENGLISH)								
Toothpick Holder (or Bud Holder)	45							
AGE HERALD (FENTON)								
Bowl, 9¼", Scarce		1,085						
Plate, 10", Scarce		1,500						
AMARYLLIS (NORTHWOOD)								
Small Compote	295	220		265			340	
Whimsey (Flattened)		375						
AMERICAN (FOSTORIA)								
Tumbler, Rare	600*							
APOTHECARY JAR								
Small Size	50							
APPLE BLOSSOMS (DUGAN)								
Bowl, 7½"	30	40		40	150		160W	
Plate, 8¼"	195	220		215	265		230W	
APPLE BLOSSOM TWIGS (DUGAN)								
Bowl	40	58	65		165		75	
Plate	200	310	320	325	675		255	
APPLE PANELS (ENGLISH)								
Creamer	30		36					
Sugar (open)	30		36					
APPLE AND PEAR INTAGLIO (NORTHWOOD)								
Bowl, 10"	75							
Bowl, 5"	30							
APPLE TREE (FENTON)								
Water Pitcher	160			430			500W	
Tumbler	44			52			140W	
Pitcher, Vase Whimsey, Rare	1,100*			1,200*				
APRIL SHOWERS (FENTON)								
Vase	40	55	60	50			100	
ARCADIA BASKETS								
Plate, 8"	50							
ARCHED FLEUR-DE-LIS (HIGBEE)								
Mug, Rare	200*							
ARCHED PANELS								
Tumbler	50							
ARCS (IMPERIAL)								
Bowls, 8½"	30	42	45				160W	
Compote	40	50					60SM	
ART DECO (ENGLISH)								
Bowl, 4"	32							
ASTERS								
Bowl, 6"	58							

Acorn

African Shield

Apple and Pear Intaglio

April Showers

228

Autumn Acorns

Banded Diamonds

Basket

Basketweave and Cable

	M	A	G	B	PO	AO	Pas	Red
ASTRAL								
Shade	45							
AUGUST FLOWERS								
Shade	36							
AURORA								
Bowl, 8½" Decorated	100	150					370IM	
AUSTRALIAN SWAN (CRYSTAL)								
Bowl, 5"	45	60						
Bowl, 9"	70	100						
AUTUMN ACORNS (FENTON)								
Bowl, 8¾"	37	45	50	40			100	1,500
Plate, Rare		1,000	950	1,100				
AZTEC (McKEE)								
Pitcher, Rare	1,300*							
Tumbler, Rare	500*							
Creamer	200						250CM	
Sugar	200						250CM	
Rosebowl							350*CM	
BABY BATHTUB (U.S. GLASS)								
Miniature Piece							75	
BABY'S BOUQUET								
Child's Plain, Scarce	90							
BAKER'S ROSETTE								
Ornament	65	80						
BALL AND SWIRL								
Mug	90							
BALLOONS (IMPERIAL)								
Cake Plate	60						90SM	
Compote	55						85SM	
Perfume Atomizer	50						85SM	
Vase, 3 Sizes	65						90SM	
BAMBI								
Powder Jar w/lid	25							
BAMBOO BIRD								
Jar, Complete		725						
BAND (DUGAN)								
Violet Hat	28	38		60				
BANDED DIAMOND AND FAN (ENGLISH)								
Toothpick Holder	65							
BANDED DIAMONDS (CRYSTAL)								
Water Pitcher, Rare	850	1,200						
Tumbler, Rare	450	350						
Bowl, 10"	90	110						
Bowl, 5"	50	65						
BANDED GRAPE (FENTON)								
Water Pitcher	210		500	400			600W	
Tumbler	37		70	45			80W	
BANDED GRAPE AND LEAF (ENGLISH)								
Water Pitcher, Rare	500*							
Tumbler, Rare	90*							
BANDED KNIFE AND FORK								
One Shape	50							
BANDED PANELS (CRYSTAL)								
Open Sugar	35	45						
BANDED PORTLAND (U.S. GLASS)								
Puff Jar	60							
BANDED RIB								
Tumbler	20							
Pitcher	120							
BARBER BOTTLE (CAMBRIDGE)								
Complete	475	650	620					
BASKET (NORTHWOOD)								
Either Shape, Ftd	85	125	160	145		590	250IG	
BASKETWEAVE (FENTON)								
Hat Shape-Advertising	35		45					
Plate (Blackberry)							500W	
Vase Whimsey, Rare	650			825				
BASKETWEAVE AND CABLE (WESTMORELAND)								
Creamer, Complete	47	70	85				210W	
Sugar, Complete	47	70	85				210W	
Syrup Whimsey	125							

229

	M	A	G	B	PO	AO	Pas	Red
BEADED								
Hatpin.............................		24						
BEADED ACANTHUS (IMPERIAL)								
Milk Pitcher.........................	75		210				150SM	
BEADED BAND AND OCTAGON								
Kerosene Lamp....................	190							
BEADED BASKET (DUGAN)								
One shape	30	50	65	58			95V	
BEADED BERRY (FENTON)								
Exterior Pattern Only								
BEADED BULLSEYE (IMPERIAL)								
Vase, 8"-14"........................	36	40	45				65SM	
BEADED CABLE (NORTHWOOD)								
Rose Bowl..........................	60	70	70	75	750*	400	600W	
Candy Dish	52	60	65	70		350	125	
BEADED HEARTS (NORTHWOOD)								
Bowl................................	47	60	60					
BEADED PANELS (IMPERIAL)								
Bowl, 8"............................	40							
Bowl, 5"............................	22							
Powder Jar w/lid	50							
BEADED PANELS (WESTMORELAND)								
Compote............................	40	50			95			
BEADED SHELL (DUGAN)								
Bowl, 9" Ftd......................	68	95						
Bowl, 5" Ftd......................	30	38						
Mug..................................	200	125		410			390W	
Covered Butter	120	175						
Covered Sugar....................	80	120						
Creamer or Spooner	75	95						
Water Pitcher.....................	460	570						
Tumbler	60	90		150				
Mug Whimsey.....................		400						
BEADED SPEARS (CRYSTAL)								
Pitcher,								
Rare	175	250						
Tumbler,								
Rare	75	90						
BEADED STARS								
Banana Boat, 8¼"...............	90							
BEADED STARS (FENTON)								
Plate, 9"...........................	90							
Bowl................................	35			75				
Rose Bowl..........................	45							
BEADED SWIRL (ENGLISH)								
Compote............................	45			55				
Covered Butter	60			70				
Milk Pitcher.......................	60			70				
Sugar	40			45				
BEADS (NORTHWOOD)								
Bowl, 8½"...........................	40	55	60					
BEADS AND BARS (U.S. GLASS)								
Spooner............................	50							
BEAUTY BUD VASE (DUGAN)								
Regular Size	34	48					55	
Twig Size, Rare	140	158						
BEAUTY BUD VASE VT.								
No Twigs...........................	28						32SM	
BEETLE ASHTRAY (ARGENTINA)								
One Size, Rare....................				500				
BEETLE HATPIN								
Complete		30						
BELLAIRE SOUVENIR (IMPERIAL)								
Bowl, Scarce......................	185							
BELLS AND BEADS (DUGAN)								
Bowl, 7½"	35	45	50	55	60			
Nappy...............................	40	65			70			
Plate, 8"............................		120						
Hat Shape	35	47						
Compote............................	40	52						
Gravy Boat, Handled	46	60			80			
BERNHEIMER (MILLERSBURG)								
Bowl, 8¼", Scarce				1,250				

Beaded Cable

Beaded Shell

Beaded Swirl

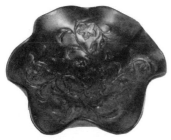

Bells and Beads

230

Bird With Grapes

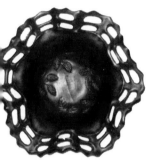

Blackberry

Blackberry Block

Blackberry Spray

	M	A	G	B	PO	AO	Pas	Red
BERRY BASKET								
One Size	45							
Matching Shakers, Pair	65							
BIG BASKETWEAVE (DUGAN)								
Vase, 6"-14"	26	40			80	180	90	
Basket, Small	35	40						
BIG CHIEF								
One Shape		85						
BIG FISH (MILLERSBURG)								
Bowl, Various Shapes	525	600	600				6,000V	
Banana Bowl, Rare		1,800	1,800					
Bowl, Tri-Cornered	1,000	1,950	1,900				1,500	
Rose Bowl, Very Rare							7,500*V	
Square Bowl, Very Rare	700	1,000	1,000				5,500*V	
BIG THISTLE (MILLERSBURG)								
Punch Bowl and Base Only, Rare		8,000						
BIRD OF PARADISE (NORTHWOOD)								
Bowl, Advertising		195						
Plate, Advertising		220						
BIRD WITH GRAPES (COCKATOO)								
Wall Vase	65							
BIRDS AND CHERRIES (FENTON)								
Bon Bon	40	60	60	60			85	
Bowl, 9½", Rare	200	325		375				
Bowl, 5", Rare	60	85						
Compote	45	60	60	60			90	
Plate, 10", Rare	1,200		1,500	1,500				
BLACK BOTTOM (FENTON)								
Candy Jar	40						60	
BLACKBERRY (FENTON)								
Open Edge Hat	37	45	50	42		150	65	480
Spittoon Whimsey, Rare	2,600			2,900				
Vase Whimsey, Rare	250			325			400W	
Plate, Rare	775			375				
BLACKBERRY (NORTHWOOD)								
Compote	52	65					95	
Bowl, 9", Ftd	47	60						
BLACKBERRY BANDED (FENTON)								
Hat Shape	30		50	40	85			
BLACKBERRY BARK								
Vase, Rare		1,500*						
BLACKBERRY BLOCK (FENTON)								
Pitcher	250	1,000	1,200	525			5,000V	
Tumbler	45	140	75	60			200W	
BLACKBERRY BRAMBLE (FENTON)								
Bowl				50				
Compote	38		60	50				
BLACKBERRY, MINIATURE (FENTON)								
Compote, Small	100	195	275	200			450W	
Stemmed Plate, Rare				375				
BLACKBERRY SPRAY (FENTON)								
Bon Bon	30	40	45	40				
Compote	35	45	50	45				
Hat Shape	30	35	40	35		195	80W	440
BLACKBERRY WREATH (MILLERSBURG)								
Plate, 10", Rare	4,000	4,700						
Plate, 6", Rare	1,490	1,600	1,700					
Bowl, 5"	50	60	65				80CM	
Bowl, 7"-9"	65	85	80					
Bowl, 10", Ice Cream	110	160		950			2,000CM	
Spittoon Whimsey, Rare			3,500					
Plate, 8" Rare			3,400					
BLOCKS AND ARCHES								
Creamer	38							
BLOCKS AND ARCHES (CRYSTAL)								
Pitcher, Rare	150	200						
Tumbler, Rare	70	85						

231

Blossoms and Band

Blossomtime

Bouquet

Brooklyn Bridge

	M	A	G	B	PO	AO	Pas	Red
BLOSSOMS AND BAND (IMPERIAL)								
Bowl, 10"............................	35	42						
Bowl, 5"..............................	20	28						
Wall Vase, Complete	42							
BLOSSOM AND SPEARS								
Plate, 8"..............................	45						40CL	
BLOSSOMTIME (NORTHWOOD)								
Compote.............................	90	180	280					
BLUEBERRY (FENTON)								
Pitcher, Scarce....................	600			850			1,000W	
Tumbler, Scarce	50			95			140W	
BO PEEP (WESTMORELAND)								
ABC Plate, Rare...................	600							
Mug, Scarce	190							
BOOT								
One Shape...........................	75							
BORDER PLANTS (DUGAN)								
Bowl, Flat, 8½"		60			170			
Bowl, Ftd, 8½"		70			185			
BOUQUET (FENTON)								
Pitcher...............................	250			475			695W	
Tumbler	35	65		50			85W	
BOUQUET AND LATTICE								
Various Shapes From $5.00-15.00 each, Late Carnival								
BOUQUET TOOTHPICK HOLDER								
One Size	65							
BOUTONNIERE (MILLERSBURG)								
Compote.............................	185	200	260				140CM	
BOW AND ENGLISH HOB (ENGLISH)								
Nut Bowl	48			55				
BOXED STAR								
One Shape, Rare							100*	
BRIAR PATCH								
Hat Shape	36	45						
BRIDLE ROSETTE								
One Shape...........................	80							
BROCADED ACORNS (FOSTORIA)								
BROCADED DAFFODILS								
BROCADED PALMS								
BROCADED ROSES								
BROCADED SUMMER GARDENS								
All Related Patterns in Similar Shapes and Colors								
Wine....................................							65	
Bon Bon							62	
Cake Tray............................							90	
Ice Bucket							90	
Covered Box							88	
Tray							90	
Cake Plate, Center Handle							95	
Vase							90	
Rose Bowl............................							90	
Bowls, Various Sizes............							65	
Flower Set, 3 Pieces.............							165	
Center Bowl, Ftd.................							100	
BROKEN ARCHES (IMPERIAL)								
Bowl, 8½"-10"	42	50	60					
Punch Bowl and Base..........	350	575						
Punch Cup	25	30						
BROCKER'S (NORTHWOOD)								
Advertising Plate.................		350						
BROOKLYN								
Bottle w/stopper..................	70	90						
BROOKLYN BRIDGE (DUGAN)								
Bowl, Scarce.......................	360							
Unlettered Bowl, Rare..........	750*							
BUBBLE BERRY								
Shade..................................							60	
BUBBLES								
Hatpin.................................		36						
Lamp Chimney							45	

232

Butterflies

Butterfly and Corn

Butterfly and Tulip

Buzz Saw

	M	A	G	B	PO	AO	Pas	Red
BUDDHA (ENGLISH)								
One Shape, Rare	875						775IB	
BULL DOG								
Paperweight	250							
BULL'S EYE (U.S. GLASS)								
Oil Lamp	185							
BULL'S EYE AND LEAVES (NORTHWOOD)								
Bowl, 8½"	37	50	50					
BULL'S EYE AND LOOP (MILLERSBURG)								
Vase, 7"-11", Rare	300	400	400					
BULL'S EYE AND SPEARHEAD								
Wine	48							
BUMBLEBEES								
Hatpin		26						
BUNNY								
Bank	30							
BUTTERFLIES (FENTON)								
Bon Bon	40	70	65	50			70	
Card Tray	35			60				
BUTTERFLIES AND BELLS (CRYSTAL)								
Compote	90	125						
BUTTERFLIES AND WARATAH (CRYSTAL)								
Compote, Large	120	250						
BUTTERFLY								
Pintray	35							
BUTTERFLY (FENTON)								
Ornament, Rare	175	200	200	200			285W	
BUTTERFLY (NORTHWOOD)								
Bon Bon, Regular	55	65	70	75				
Bon Bon, Ribbed Exterior		260					425IB	
BUTTERFLY (U.S. GLASS)								
Tumbler, Rare	5,500							
BUTTERFLY AND BERRY (FENTON)								
Bowl, 10", Ftd	90	200	240	210			375W	
Bowl, 5", Ftd	30	40	50	40			85	1,100
Covered Butter	130	260	300	250				
Covered Sugar	100	170	190	150				
Covered Creamer	100	170	190	150				
Nut Bowl Whimsey		750		750				
Spooner	100	160	185	150				
Pitcher	375	685		620			1,100W	
Tumbler	45	75	65	60			90W	
Hatpin Holder, Rare	750			620				
Vase, Rare	55			195			400	650
Spittoon Whimsey, 2 Types		2,500		2,500				
Bowl Whimsey (Fernery)	800	1,100		1,350				
Plate, Ftd (Whimsey)				1,350				
BUTTERFLY BOWER (CRYSTAL)								
Compote	80	115						
Cake Plate, Stemmed		175						
BUTTERFLY BUSH (CRYSTAL)								
Compote, Large	110	150						
BUTTERFLY AND CORN (NORTHWOOD)								
Vase, Rare							3,000V	
BUTTERFLY AND FERN (FENTON)								
Pitcher	400	515	575	600				
Tumbler	38	47	48	50				
BUTTERFLY AND TULIP (DUGAN)								
Bowl, 10½", Ftd, Scarce	475	2,000						
Bowl, Whimsey Shape, Rare	850	1,250						
BUTTERMILK, PLAIN (FENTON)								
Goblet	50	65	70					
BUTTON AND FAN								
Hatpin		55						
BUTTONS AND DAISY (IMPERIAL)								
Hat (Old Only)							60CM	
Slipper (Old Only)							70CM	
BUTTRESS (U.S. GLASS)								
Pitcher, Rare	300							
BUZZ SAW								
Shade	40							
BUZZ SAW (CAMBRIDGE)								
Cruet, small 4", Rare			400					
Cruet, Large 6", Rare	400		400					

Cane and Scroll

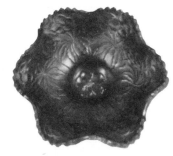

Carolina Dogwood

Cathedral

Chatelaine

	M	A	G	B	PO	AO	Pas	Re
CACTUS (MILLERSBURG)								
Exterior Pattern								
CANADA DRY								
Bottle..................................	14						28W	
CANDLE LAMP (FOSTORIA)								
One Size	90						120AM	
CANE (IMPERIAL)								
Bowl, 7½"-10"	30						45	
Pickle Dish	25						45	
Wine	50						60	
Compote..................................	70						90	
CANE AND DAISY CUT (JENKINS)								
Vase	90							
Basket, Handled, Rare..........	175						190	
CANE AND SCROLL (SEA THISTLE) (ENGLISH)								
Rose Bowl...........................	55			70				
Creamer or Sugar	45							
CANNONBALL VT.								
Pitcher.................................	250			300			400W	
Tumbler	60			70			85W	
CANOE (U.S. GLASS)								
One Size	75							
CAPITOL (WESTMORELAND)								
Mug, Small.........................	75							
Bowl, Ftd, Small		65		65				
CAPTIVE ROSE (FENTON)								
Bowl, 8½"-10"	37	50	48	45				
Compote..............................	40	60	55	48				
Plate, 7"...............................	80	110	100	95				
Plate, 9"...............................	150		325	265				
Bon Bon	40		60	55				
CARNATION (NEW MARTINSVILLE)								
Punch Cup	45							
CARNIVAL BELL								
One Size	350							
CARNIVAL HONEYCOMB (IMPERIAL)								
Bon Bon	35	45	45					
Creamer or Sugar	28							
Plate, 7"...............................		80						
Bowl, 6" Handled	30							
CAROLINA DOGWOOD (WESTMORELAND)								
Bowl, 8½"	65	90			225MO	450		
Plate, Rare...........................					290MO			
CAROLINE (DUGAN)								
Bowl, 7"-10"	52				185			
Banana Bowl					300			
Basket, Scarce.....................					400			
CARRIE (ANCHOR-HOCKING)								
One Size	55							
CARTWHEEL #411 (HEISEY)								
Compote	40							
Goblet	60							
CATHEDRAL (SWEDEN)								
(Also know as Curved Star)								
Chalice, 7"...........................	75			115				
Pitcher, Rare.......................				2,500				
Bowl, 10"..............................	40			50				
Flower Holder	60							
Epergne, Scarce...................	425							
Compote, 2 Sizes	40			55				
Creamer, Ftd	45							
Butterdish, 2 Sizes	185							
CATHEDRAL ARCHES (ENGLISH)								
Punch Bowl, 1 Piece	250							
CATTAILS								
Hatpin...................................		26						
CENTRAL SHOE STORE (NORTHWOOD)								
Bowl, 6"-7"		285						
CHAIN AND STAR (FOSTORIA)								
Tumbler	750							
Covered Butter, Rare	900*							
CHATELAINE (IMPERIAL)								
Pitcher, Rare.......................		2,800						
Tumbler, Rare		460						
CHATHAM (U.S. GLASS)								
Compote	65							
Candlesticks, Pr	75							

Tracy Luther
AUCTIONS & ANTIQUES

Full Service
Auction & Estate
Services
(651) 770-6175
(Fax) 770-6906
www.lutherauctions.com

Finer Antique & Art
Auction Catalog for July 25, 2005
Preview: Sunday 5 pm - 8 pm & Monday 10 am – 6 pm
Auction: Monday, 6 pm

A 10% COMMISSION WILL BE CHARGED TO THE BUYER ON ALL SALES.

Thank you for coming. Please take the time to carefully look over all items of interest to you. The purpose of this catalog is only for the identification of merchandise. It is our intention to give an accurate description to the best of our knowledge and note any defects we may be aware of. However, the Auctioneer makes no guarantee or warranty, expressed or implied, as to the accuracy of the information given on any merchandise found in this catalog. Everything is sold AS IS, WHERE IS. It is the sole responsibility of you, the buyer, to judge the value, condition and authenticity of merchandise and bid accordingly. We do not guarantee the authenticity of any signature. We do not guarantee the value, condition or authenticity or any item. Everything is sold "as is". We disclaim responsibility for any opinions expressed by the auctioneer or others as to value of any item. Please use the preview times to inspect all items before bidding on them. Note: items marked with a (R) are being sold subject to a reserve. The reserve will not exceed the low end of the printed auction estimate. *All sales are final.*

All items must be removed from the premises by 5:00PM on Tuesday. If you purchase an item and are unable to pick it up by Tuesday, make arrangements with the Auctioneer prior to this time.

Consignments are taken at 2556 East 7th Avenue, North St Paul daily by appointment. Appointments welcomed anytime, call (651) 770-6175.

We appreciate your business. If you have any questions or comments please bring them to the Auctioneer or any available staff member. Thank you.

Please note upcoming auctions
Antique, Estate & Collectibles - August 1st
Antique, Estate & Collectibles - August 8th
Antique, Estate & Collectibles - August 15th

1. 6" purple Peacock & Urn bowl & Marigold compote...x2
2. Northwood 6 1/2" Lattice Twigs white shallow bowl
3. 3 pieces Imperial Grape- bowls...x3
4. 2 Imperial Heavy Grape bowls...x2
5. 3 Fenton vintage bowls...x3
6. Fenton Persian Medallion orange bowl
7. Grape & Cable 9" bowl
8. Grape & Cable 9" plate with basketweave back
9. Northwood Grape Leaves bowl
10. Peacock at Fountain 5 piece water set...lot
11. Fenton Heart & Vine blue bowl
12. Fenton Heart & Vine green bowl
13. Northwood powder jar- Grape & Cable
14. Grape & Cable bon bon- Pas Marigold
15. Grape & Cable 2 sides up hat
16. 3 pieces Northwood Grape & Cable...x3
17. 2 Imperial Grape & 2 Holly compotes...x4
18. 2 marigold Grape & Cable spoons, marigold sugar lid & green Grape & Cable covered sugar...x4
19. Imperial Windmill dresser tray & 2 bowls...x3
20. Fenton Thistle banana boat- cobalt blue
21. Windmill & Mums center bowl
22. Fenton Grape & Cable Spat footed bowl & Windmill & Mums marigold bowl...x2
23. 4 Fenton Acorn bowls- moonstone, marigold & 2 blue...x4
24. Fenton Butterfly & Berry 7 piece water set...lot
25. Imperial Grape green 6 piece water set...lot
26. Fenton enameled Cherry Blossom 7 piece water set...lot
27. 2 Peacock at the Fountain blue tumblers...x2
28. Cathedral Square berry bowl- 7 3/4"
29. 9 pieces Rays- 6 footed sherbets, bowl, plate, & tray...lot
30. Grape & Cable 8" footed bowl- green
31. 11" Stag & Holly footed bowl- amethyst
32. 10" Stag & Holly footed bowl- marigold
33. 9" Northwood Good Luck bowl- marigold
34. 2- 6 1/2" Northwood Drapery candy dishes- purple & white...x2
35. 8" Floral & Wheat handled bon bon- white
36. 8" Star of David & Bows Grape Northwood bowl- purple
37. Child's Kittens cup- marigold
38. Stork & Rushes Northwood Marigold covered butter & creamer...x2
39. 9 1/2" Little Fishes Fenton footed bowl- cobalt
40. 2 Panther 5 1/2" footed bowls- marigold...x2

41. Stork & Rushes punch cup & bowl & Cockatoo wall pocket...x3
42. Fenton Dragon & Lotus bowl- blue
43. Fenton Dragon & Lotus bowl- marigold
44. Fenton Dragon & Lotus bowl- amethyst
45. Butterfly & Berry 6 piece berry set- marigold...lot
46. 3 pieces Butterfly & Berry- covered butter, covered sugar & spooner...x3
47. Blue Butterfly & Berry creamer
48. Fenton amethyst Butterfly bon bon & Northwood Butterfly bon bon...x2
49. Dugan pony bowl- marigold
50. 4 Fenton enameled Prism Band tumblers...x4
51. 2 insulators...x2
52. 2 marigold Tree Bark pitchers, 6 tumblers & covered candy...lot
53. 10" Fenton Cherry Chain bowl- marigold
54. Purple Imperial Jewels cuspidor shape vase
55. 2 goblets- blue Thistle Panel & Stippled Rose...x2
56. Orange Tree covered powder jar- blue
57. 4 pieces marigold Orange Tree (2 as seen)...x4
58. Millersburg Primrose bowl- marigold
59. 3 marigold candlesticks, rose bowl, & Leaf Rays nappy...x5
60. 5 1/2" Birds & Cherries compote- green
61. 12" Daisey Block rowboat- marigold
62. 4 pieces marigold English Carnival (2 as seen)...x4
63. Sowerby covered hen- marigold
64. 13" Cherry Wreath banana boat- amethyst
65. 6" Cherry Wreath berry bowl- blue
66. Ribbed panel Jack in Pulpit style epergne with metal base- clear to marigold
67. Ribbed panel jam/mustard pot in metal stand- vaseline/marigold iridescent
68. 5 pieces Stippled Rays- 2 green, 1 amethyst, & 2 marigold...x5
69. Northwood Stippled Scales compote- green
70. 5 piece Octagon water set- marigold (pitcher as seen)...lot
71. 2 marigold vases & marigold rose bowl...x3
72. 6 pieces marigold- 2 Butterfly trays, Clover Leaf tray & 3 goblets...x6
73. 4 Pansy sugar/creamers- 2 marigold, 1 amethyst, 1 amber...x4
74. 6 pieces misc. carnival glass- Rays, Prisms, etc...x6
75. 10 pieces depression era carnival glass...lot
76. 10 piece Westmoreland Orange Peel punch set (1 cup as seen)...lot
77. Imperial Long Hobstar punch bowl & base...lot

78 Orange Tree bowl, base (crack) & 1 cup...lot   _____

79 9 assorted punch cups including Acorn Burr, Wreath of Roses, Luster Flute, etc & Memphis punch bowl base...x10   _____

80 3 depression era glass figural powder jars & 3 figural banks...x6   _____

81 Northwood Peacock at the Fountain master berry bowl- amethyst   _____

82 Purple & blue iridescent Fruits & Flowers 2 handled bon bon with basketweave   _____

83 Marigold Northwood Fruits & Flowers 2 handled bon bon with basketweave   _____

84 Fenton amber Strawberry Spray 2 handled nappy   _____

85 Fenton Daisey Pinwheels & Beaded Cable 8" bowl- marigold   _____

86 3 Carnival Holly 6" & 6 1/2" hat shape bowls- cobalt, amber, etc...x3   _____

87 Purple Northwood Fruits & Flowers 2 handled bon bon with basketweave   _____

88 5 pieces misc carnival glass- Octagon, Bulls Eye, Stippled Rays, etc...x5   _____

89 6 pieces assorted small carnival glass- swan, shell, basket, etc...x6   _____

90 Fenton white Orange Tree dish with Beaded Berry back   _____

91 Fenton blue Orange Tree dish with Beaded Berry back   _____

92 Purple Grape & Cable sweet meat compote   _____

93 Fenton Floral & Grape pitcher- marigold   _____

94 5 blue paneled tumblers...x5   _____

95 Imperial octagon 7 piece wine set- marigold...lot   _____

96 Imperial Grape 7 piece wine set- marigold...lot   _____

97 2 Fenton Peacock Tail hat shape bowls with advertising...x2   _____

98 Fenton Blackberry blue miniature compote   _____

99 Fenton Peacock Tail hat shape bowl & double handled nappy...x2   _____

100 Palm Beach vase- 5 1/2"   _____

100A Framed signed Bruce Lattig watercolor- ducks 21"x 29"   _____

100B Dali portfolio with 40 lithos...lot   _____

100C Group of assorted unframed wildlife, etc art...lot   _____

100D Recast Remington bronze "Bronco Buster" 23"   _____

101 8 1/2" Concord bowl- marigold   _____

102 Imperial Jewels rose bowl- purple (Iron Cross mark)   _____

103 2 Coin Spot bowls & Frosted Block bowl...x3   _____

104 Fenton 9" emerald green Holly bowl   _____

105 Fenton 9" green Holly bowl- ribbon edge   _____

106 Maple Leaf footed berry bowl & Rosette 3 footed bowl...x2   _____

107 Northwood bushel basket- blue   _____

108 2 Lotus & Grape footed bowls- marigold & blue...x2   _____

109 9" Purple Blackberry Wreath shallow bowl   _____

110 2 Luster Rose ferneries- blue & smoke...x2   _____

111 Fenton Holly bowl & Persian Medallion 2 handled nappy...x2   _____

112 Star medallion & Poinsettia pitcher...x2   _____

113 Northwood Beaded Cable rose bowl- purple   _70_

114 Dugan 6" scales bowl with panels & beading & Fenton double stem rose bowl...x2   _____

115 10" Diamond Prisms footed cake plate   _____

116 Hobstar & Arches 2 piece fruit bowl, Diamond & Fine Rib dish & white Octagon bowl...x3   _____

117 Northwood peach opalescent Stippled Petal fruit bowl with enamel   _____

118 Jeweled Heart peach opalescent master berry & berry bowl...x2   _____

119 3 pieces peach opalescent- Stippled Cosmos, Stippled Petal & Shell. & Fan gravy...x3   _____

120 Millersburg green Holly 9" bowl   _____

121 3 pieces Grape Delight- 2 rose bowls & nut dish...x3   _____

122 Purple Dahlia creamer with heavy scroll feet (feet rough)   _____

123 Dugan 9" purple Flowers & Frames footed bowl   _____

124 Millersburg 7" Holly 2 handled nappy- amethyst   _____

125 Imperial 7" Golden Honeycomb purple candy dish   _____

126 Aqua Opal Hearts & Flowers compote   _____

127 Grape decanter (no stopper) & 2 Golden Harvest wines...lot   _____

128 5" Dugan marigold beaded basket & 9" Fenton Stippled Rays green bowl...x2   _____

129 2 Imperial rippled marigold vases (approx. 16")...x2   _____

130 2 Northwood 10 3/4" Plume panel vases- marigold...x2   _____

131 3 Imperial rippled marigold vases- 10" & 2- 11 1/2"...x3   _____

132 8 1/4" Fenton green Rib vase & 9 1/2" Westmorland teal vase...x2   _____

133 2 marigold Jack in the Pulpit vases- 9 1/2" & 6"...x2   _____

134 3 assorted carnival vases- 2 Diamond Point & 1 Diamond Rib...x3   _____

135 6 1/2" Rib & Panel bud vase & 5 1/2" Ripple vase- marigold...x2   _____

136 7 1/4" cobalt vase & 10 1/2" marigold Hobnail vase...x2   _____

137 Fenton Blackberry Spray hat shape vase- red   _____

138	7" Cambridge Inverted Strawberry bowl- "Near Cut"	
139	8 3/4" Northwood 3 Fruits plate- amethyst	
140	Marigold Orange Tree hat pin holder	
141	4 hat pins including 6 plums, etc...x4	
142	4 misc pieces Marigold carnival including Fine Cut & Star, Kokomo, etc...x4	
143	11" Fenton Wild Daisey & Lotus 3 footed bowl- Marigold	
144	7 1/4" Fenton Wild Daisey & Lotus Spatula footed bowl-Marigold	
145	9" Imperial Open Rose plate- Amber	
146	Fenton Peacock & Urn beaded berry- Cobalt	
147	Fenton Ten Mums bowl- Marigold	
148	Imperial 11 1/2" wide Open Rose 3 footed bowl- Smoke	
149	8 3/4" Carolina Dogwood bowl- Marigold	
150	7" wide Soda Gold cuspidor & 2- 7 1/2" candlesticks...x3	
151	10 3/4" & 9 1/4" Marigold Open Rose bowls...x2	
152	4 pieces misc carnival including Blackberry Banded, etc (1 with chips)...x4	
153	3 1/2" Cobalt Singing Birds mug *260*	
154	Dugan Purple Holly & Berry 1 handle bon bon	
155	2 carnival hat shaped vases & 2 Marigold mugs...x4	
156	3 assorted carnival rose bowls- Garland, Louisa, Daisey & Plum...x3	
157	4 misc pieces carnival glass- bowls, etc including Hottie bowl, Thistle & Thorn, etc...x4	
158	3 open edge basket weave bowls including Blackberry...x3	
159	8 assorted Marigold carnival tumblers including Field Thistle, Grape & Gothic, etc (some as is)...lot	
160	Northwood Grape Delight Fernery & Marigold Sailboat compote...x2	
161	Grape & Cable hat pin holder	
162	Group of misc carnival glass- bowls, etc (approx 21 pieces)...lot	
163	Set of Marigold Bouquet & Lattice dinnerware (approx 68 pieces)...lot	
164	6 misc creamers & sugars (no lids) including Shell & Jewel, Stipped Rays, etc...x6	
165	Purple Imperial Grape 7 piece wine set (stopper as is)...lot	
166	Amethyst Mary Ann loving cup *475 footed*	
167	Amethyst Millersburg Fleur de lis bowl *250*	
168	Northwood 10" Peacock & Urn Ice Green bowl	
169	Northwood Pastel/Ice Green corn vase	
170	Aqua Opalescent Blue Beaded Cable rose bowl 4 1/2" wide	
171	Northwood Ice Blue Opalescent Pillar & Drape rose bowl, 5 1/2" wide	
172	Fancy carved mahogany 2 door china cabinet with mirrored back (glass as seen)	
173	Fancy carved oak curved glass china cabinet with lions heads & claw feet	
174	Oak framed L.M. Roth litho- Jesus with Children (35"x 48")	
175	Framed German litho- Guardian Angel (21 1/2"x 32")	
176	Set of Friendly Village dinnerware- over 70 pieces...lot	
177	9 1/2" art glass bowl & 7 1/2" art pottery bowl...x2	
178	12 1/2" & 10" vaseline opalescent Hobnail baskets...x2	
179	5 pieces opalescent glass- cream/sugars etc...x5	
180	Collection of assorted porcelain, etc birds including slag hen on nest...lot	
181	Ingraham oak naval motif shelf clock	
182	4 pieces assorted Victorian, custard glass, etc including Mary Gregory tumbler...x4	
183	Large group of assorted carnival glass including punch set, etc on shelf...lot	
184	Group of carnival glass toothpicks, figural covered nests, shakers, etc on shelf...lot	
185	Carved & painted French style 2 door chest with marble top	
186	Carved & painted French style curio cabinet- 70" h.	
187	Carved & painted French style curio cabinet- 70" h.	
188	30" sea shell mirror & 18" sea shell wreath...x2	
189	Collection of pottery & porcelain figures including dolls with glass dome...lot	
190	Set of rose pattern dinnerware including covered butter (approx. 28 pieces)...lot	
191	6 pieces Fenton, etc. carnival glass including opalescent Drapery bowls...x6	
192	2 large shadow box mirrors with shelves...x2	
193	Empire style china cabinet with curved glass door	
194	2 framed Martin Luther lithos dated 1882...x2	
195	Collection of 25 figural liquor bottles...lot	
196	Collection of 26 figural liquor bottles...lot	
197	3 porcelain decorated tea sets including hand painted...x3	
198	Large group of colored pattern, etc glass including Claret jug...lot	
199	Group of assorted pottery & porcelain on shelf including Frankoma, Lefton, etc...lot	
200	Framed Odd Fellows litho by Strobidge & Co dated 1870	
201	Framed vintage litho- Chariot (10"x 19 1/2")	

202	Framed signed oil on canvas (24"x 36") & still life print...lot
203	12 Franklin Mint pewter figures- Men & Women who Built America...x12
204	22 assorted Fenton carnival glass figures...x22
205	Framed signed Baroni oil on canvas- castle (24"x 36")
206	Pine narrow kitchen cabinet
207	Painted Deco kitchen cabinet (as seen)
208	2 inlaid Oriental plaques...x2
209	Signed Pleaton oil on canvas- cabin lake scene (24"x 48")
210	2- 20" bronze cranes...x2
211	Antique handmade Persian Oriental rug (approx. 4'x 6')
212	Antique handmade Persian Oriental rug (approx. 4'x 6')
213	2 antique handmade Oriental rugs (approx. 4'x 6' & 5'x 7' worn)...x2
214	2 small antique handmade Oriental rugs (worn)...x2
215	2- 8 1/2" marble busts of ladies...x2
216	2 miniature souvenir jugs...x2
217	Victorian double inkwell
218	Signed oil on canvas (6"x 9") & hand painted tile (6"x 6")...x2
219	Vintage mandolin & uke...x2
220	6 perfumes & 11 glass shoes...x17
221	10 unframed illustration artworks- Fenelle...x10
222	Stereo viewer & group of over 100 view cards...lot
223	25 bone china cup & saucer sets...x25
224	Slate & bronze mantel clock (as seen)
225	7 1/2" Lundberg studios art glass vase
226	Silver-plate Swedish soup ladle
227	28 pieces American Classic pattern sterling flatware...x28
228	18 Swedish sterling spoons...x18
229	11 Norway, Danish sterling spoons...x11
230	10 sterling pins- Danecraft, etc...x10
231	2 Jadeite boudoir lamps...x2
232	Vintage Deco figural globe lamp
233	Deco figural ashtray & cigarette box stand
234	2 porcelain Deco bird candle holders...x2
235	11" Oriental bronze vase
236	Seth Thomas mantel clock
237	Victorian spoon carved writing table
238	Victorian walnut marble topped chest with mirror
239	Wooden log cabin dollhouse with furniture...lot
240	10 pieces 1920's Deco glass...x10
241	Group of Wileman Pre-Shelley porcelain on shelf...lot
242	Antique claw foot triple curve glass china
243	30" vintage mechanical Santa & bisque collector doll...lot
244	Wainscote store counter 77"x 34"
245	Handmade Oriental rug (approx. 9'x 12')
246	10K ring, 14K ring & 23K floating paperweight...x3
247	Arts & Crafts woodblock signed Bengtsson & Wainwright linocut...x2
248	WPA painting of street worker signed Bob Brown (24"x 30")
249	2 modernist paintings (1 signed)...x2
250	Mexican Arts & Crafts carved wood table with glass top
251	Stickley Bros. mission oak telephone table with label
252	Red Wing Tampico water cooler (complete)
253	Large set of Red Wing Tampico dinnerware including serving pieces...lot
254	Set of Red Wing Bob White dinnerware including serving pieces...lot
255	Victorian hanging fixture
256	Set of Kawai drums
257	"Triple" 4 string banjo
258	Approx. 40 pieces rose pattern, etc glassware...lot
259	10" Roseville Fuschia bowl
260	7" Roseville Zephyr Lily basket (base chip)
261	8" Roseville Pinecone vase (base chip)
262	9" Roseville Foxglove vase
263	10" Roseville Pinecone vase (chip repair)
264	Branded Limbert high back rocker
265	Signed & numbered Joseph Hirsch litho (22"x 27")
266	Signed Martin DeWitt acrylic on paper (3 1/2"x 9")
267	Stickley Bros. footstool with brass tag
268	Carved Oriental blanket chest
269	Mission oak psychiatrists couch
270	Deco green glass lamp with vintage Mica shade
271	Art glass vase in Art Nouveau stand 10 3/4" high
272	13" hand painted Limoges vase
273	Arts & Crafts silver-plate match holder with enamel plaque
274	5" signed Saturday Evening Girls pottery teapot (small chip)
275	2- 9 1/2" fine Cloisonné vases (1 as seen)...x2
276	7 early decorated Oriental pottery bowls...x7
277	Carved & upholstered Victorian loveseat
278	Victorian oak platform rocker
279	6 light brass & beveled glass chandelier
280	6 carved & upholstered Queen Anne style chairs...x6
281	35" gilt oval mirror

282   35" fancy gilt mirror with shell pediment   _____

283   39" ornate gilt mirror   _____

284   Sterling & turquoise bracelet & silver Naja pendant...lot   _____

285   Group of assorted sterling, etc jewelry, carved bone fetish necklace, etc...lot   _____

286   14K hinged bangle bracelet set with 7 opals & 13 diamonds (42.3 grams)   _____

287   14K ring set with baguette & round diamonds (approx. 1.00 cwt)   _____

288   10K necklace channel set with banquette diamonds (approx. 1.00 cwt)   _____

289   14K ring set with rubies & diamonds   _____

290   14K gold ring set with 18 diamonds & red stone   _____

291   14K gold ring set with tanzanite & diamonds (chips)   _____

292   Gerard Perregaux watch in 14K case   _____

293   Group of vintage jewelry including 14K, Deco, sterling, charm bracelet, etc...lot   _____

294   2 music cabinets - 1 oak,,,x2   _____

295   Painted corner shelf, 12 drawer cabinet & painted Victorian 4 drawer chest...x3   _____

296   Lane Queen Anne style mahogany cedar chest   _____

297   Group of assorted sterling articles including vases, compotes, etc...lot   _____

298   Large set of Limoges china...lot   _____

299   Group stag handled, etc carving sets...lot   _____

300   10 volumes History Civil War- 1911 (as seen)...x10   _____

301   Large volume Atlas City of Mpls (as seen)   _____

302   Early stereo slide projector   _____

303   F.O.K. catalog, store tin & Curtiss peanut jar...x3   _____

304   Lufkin advertising box & hardware display...x2   _____

305   Collection of oil company etc tins & cigar boxes...lot   _____

306   Eberhard pencil display & 2 sharpeners...x3   _____

307   Machinists chest with oak interior   _____

308   High grade carved 6 1/2' mahogany sideboard   _____

309   Plaster classical frieze approx. 4' long (as seen)   _____

310   Fenton Apple Tree marigold apple pitcher   _____

311   Framed oil on canvas- children (19 1/2"x 25 1/2")   _____

312   2 signed C.J. Fox oil portraits (25"x 30")...x2   _____

313   Stacking mahogany bookcase/desk   _____

314   4 piece inlaid satin wood French style bedroom suite...lot   _____

315   Handmade Oriental rug- animal design (approx. 8'x 10')   _____

316   13" Goebel W German elephant figure   _____

317   2 Wallendorf 9 1/2" dancer figures...x2   _____

318   Early Goebel crown mark figure of a girl & figural vase...x2   _____

319   Early Goebel hat pin holder & Goebel shaker...x2   _____

320   Goebel Disney Snow White with 9 dwarfs- full bee mark...lot   _____

321   7 Goebel Disney, etc figures...x7   _____

322   7 Goebel Disney, etc figures...x7   _____

323   2 Goebel Ducal Crown hair receivers & 2 powder boxes...x4   _____

324   Early Goebel triangle & moon mark blue & white pitcher 4"   _____

325   4 Goebel Ducal Crown blue & white 4" pitchers...x4   _____

326   3 Goebel Ducal crown pieces...x3   _____

327   2 Goebel crown mark figural perfumes...x2   _____

328   Antique American cherry wood 4 drawer chest   _____

329   Cherry 3 drawer chest with mirror & bed with rails...lot   _____

330   Cherry blanket chest   _____

331   14 pieces Fenton, etc glass...x14   _____

332   Large framed Parrish print   _____

333   Framed signed Redlin print "Sharing Seasons"   _____

334   8" Roseville Zephyr Lily vase (repair)   _____

335   2 Roseville 6" Bushberry ewers...x2   _____

336   10" Roseville Snowberry planter box   _____

337   7" Roseville vase   _____

338   2 Weller berry pattern handled vases...x2   _____

339   6" Weller berry pattern jardiniere   _____

340   Roseville 3 piece console set...lot   _____

341   14K white gold ring set with large amethyst & several diamonds .50 cwt   _____

342   Hallmarked gold bead necklace with amethyst & enamel clasp   _____

343   Pair 14K white gold chandelier earrings set with amethyst & diamonds   _____

344   14K white gold filigree bracelet set with many diamonds 1.30 cwt   _____

345   14K white gold filigree pin set with several diamonds .50 cwt   _____

346   Antique melodeon in rosewood case   _____

347   Antique Empire flame mahogany sideboard   _____

348   Carved mahogany Victorian sideboard   _____

349   Mahogany 6 drawer chest   _____

350   Framed Harper's poster 12"x 18 1/2"   _____

351   Framed Harper's Edward Penfield poster 11"x 15"   _____

352   Framed Harper's Edward Penfield poster 11"x 16" (as seen)   _____

353   Framed signed Schofield oil on canvas- river landscape 40"x 33"   _____

354   Group of early photos, framed portrait, etc...lot   _____

355   Group of Harpers, lithos, silk painting, etc...lot   _____

356	Harley Davidson oil can & Polarine oil can (as seen)...x2	_____
357	2 Goebel Disney Bambi trays...x2	_____
358	Goebel Disney Bambi figure & Bambi vase- full bee...x2	_____
359	5 Goebel Disney Bambi figures...x5	_____
360	Goebel Disney cat figure- full bee	_____
361	2 early Goebel Kewpie figural perfumes...x2	_____
362	Early Goebel Deco perfume & dog perfume...x2	_____
363	2 Goebel Ducal Crown figural match holders (hairline)...x2	_____
364	10" Wallendorf figure- nude with deer	_____
365	5 porcelain & 2 crystal Goebel fish figures...x7	_____
366	3 early Goebel Ducal crown figures- children (1 as seen)...x3	_____
367	Early Goebel plaque of Napoleon	_____
368	Mahogany table with 8 chairs, 6 leaves & buffet...lot	_____
369	Flame mahogany small server	_____
370	Flame mahogany china cabinet	_____
371	Ornate beveled mirror 22"x 48"	_____
372	14K men's ring set with 6 baguette diamonds .60 cwt	_____
373	14K ring set with marquise sapphire & 18 diamonds	_____
374	14K bracelet set with 40 diamonds .75 cwt, 14 gr	_____
375	14K ring set with 3 garnets & 6 diamonds	_____
376	10K ring set with several garnets	_____
377	14K pendant set with diamonds & red stone on 14K chain	_____
378	Decorated round oak table	_____
379	Stained & leaded glass window 24"x 36" (some damage)	_____
380	Rosewood writing desk with carved legs	_____
381	Oak cabinet with leaded glass doors	_____
382	2 Goebel Schaubach figures- boy with goat & boy with duck...x2	_____
383	2 Wallendorf figures- boys with dogs...x2	_____
384	2 Goebel Schaubach figures- boy with sheep & boy with deer...x2	_____
385A	Round oak pedestal table	
385	8" Schaubach Kunst figure- nude on horse back	_____
386	10" Schaubach Kunst figure- lady with dog	_____
387	4- 5" Ducal Crown, etc half dolls...x4	_____
388	2- 6" Ducal Crown half dolls...x2	_____
389	2 early Goebel crown mark half doll lamps...x2	_____
390	5" Ducal Crown half doll- girl with tea set	_____
391	Ducal Crown blue & white chamber stick	_____
392	Deco oak wardrobe with beveled mirror	_____
393	Heavy pine trunk, stand & table with drawers...x3	_____
394	Cedar chest	_____
395	2 reverse paintings in convex oval frames...x2	_____
396	2 armchairs...x2	_____
397	Mission oak hall bench with beveled mirror	_____
398	Pressed oak tabouret stand	_____
399	Pair of 14K white gold chandelier earrings set with diamonds .50 cwt	_____
400	10K white gold ring set with blue topaz & 2 diamonds	_____
401	Antique Chippendale style writing table	_____
402	Rosewood console table with drawer	_____
403	Antique Empire mirror 30"x 48"	_____
404	Pencil signed numbered J D Knap litho 22"x 18" "Reflections"	_____
405	Sterling Versaille pattern cake server	_____
406	4 sterling Versaille pattern serving pieces...x4	_____
407	Sterling Versaille pattern carving set...lot	_____
408	12 sterling Versaille pattern small spoons...x12	_____
409	52 pieces assorted sterling Versaille pattern flatware...x52	_____
410	12 carved mother of pearl & silver knives...x12	_____
411	17 pieces sterling flatware...x17	_____
412	4 sterling candlesticks...x4	_____
413	Signed Friedell Pasadena sterling dresser frame 14"x 10"	_____
414	Sheffield punch bowl, 2 candelabra & sterling basket frame...lot	_____
415	10 cut goblets, music box, knife rest, misc glasses, etc...lot	_____
416	Early St Paul & Yale framed prints 9 1/2"x 7"...x2	_____
417	Book "Complete Angler" 1835 full leather	_____
418	Mahogany double pedestal table with 10 shield back chairs & 3 leaves...lot	_____
419	Kensington inlaid Hepplewhite style server	_____
420	Mahogany & gilt wood mirror 19"x 45" (some damage)	_____
421	4 vols. "Shores of the Polar Sea" by J Franklin 1829 (full leather)...x4	_____
422	5 vols. "Beacon of Lights of History" 1886, 3/4 leather...x5	_____
423	2 vols. Memoirs Count Grammont 1903 3/4 leather...x2	_____
424	14 vols. Oliver Wendell Holmes 1900, 3/4 leather...x14	_____
425	Book- W Sommerville Poems 1727 (as seen)	_____
426	3 leather bound books "Life a Bee", "Marcus Aurelius", etc...x3	_____
427	Book "Works of Hogarth" illustrated full leather	_____
428	4 vols. including Dickens, Cambridge poems, etc- leather...x4	_____
429	29 leather bound books (as seen)...lot	_____
430	Antique claw foot American 9 drawer chest	_____

431	Cedar chest
432	39" mahogany pedestal drum table
433	Upholstered chaise lounge
434	Framed oil on board- Russian man fishing, 7"x 9"
435	Framed oil on board- Russian man with gun & dog 7"x 9"
436	Bob Brown oil on board 10"x 12" signed on reverse
437	Signed G Resler oil on board- landscape with trees 9"x 12"
438	Unframed oil on canvas- blacksmith 17"x 14"
439	Framed signed E E Claridge- baseball monkey 18 1/2"x 22 1/2"
440	Framed signed E E Claridge- Monkey Party 18 1/2"x 22 1/2"
441	14K white gold ring set with English blue topaz & several diamonds, 17 gr
442	14K men's ring set with 10 diamonds 1.0 cwt
443	88" signed Nutting pine bench Windsor style
444	Pine Windsor style chair
445	Wooden tool box with tools
446	Brass fireplace fender & tools
447	14K white gold ring set with several princess & baguette diamonds 2.0 cwt
448	14K chandelier pendant on chain with diamonds 1.37 cwt
449	3 Goebel Nasha napkin figures...x3
450	Goebel Ducal Crown half doll 6"
451	4- 3" Goebel half dolls- crown marks, etc...x4
452	6 Goebel crown mark plaid children figures...x6
453	5 miniature Goebel half dolls...x5
454	3- 4 1/2" Goebel crown mark dolls...x3
455	Goebel Ducal Crown half doll with tea set 4"
456	7 Goebel novelty items including toothbrush holder, etc...x7
457	Antique handmade Sarouk Oriental rug (approx 9'x 12')
458	Handmade Persian Oriental rug (approx 8'x 10')
459	Handmade Oriental rug (approx 7'x 10')
460	Handmade Oriental rug (approx 7'x 10')
461	Handmade Oriental rug (approx 5'x 6')
462	Steuben blue Aurene over yellow jade etched lamp
463	Unsigned Steuben etched calcite shade
464	2 hand painted & gilt 14" vases...x2
465	2 hand painted & gilt 18" vases...x2
466	9" hand painted hinged jewel box
467	5 piece iron patio set...lot
468	Antique iron patio chair (as is)
469	2 framed contemporary oils by Dana Lipsig...x2
470	Framed signed mixed media artwork- horse race 18"x 24"
471	Framed Victorian needlework item "Walk in Love"
472	Oak 2 door claw foot bookcase with leaded doors
473	5' mission oak built-in buffet (as seen)
474	26" ornate oval mirror
475	Framed Picasso litho, New York Graphic Society label 20"x 29"
476	Unframed signed oil on canvas 24"x 33"
477	Framed mission oak beveled mirror 26"x 48"
478	Framed Victorian walnut mirror 16"x 28"
479	Victorian marble top plantstand
480	Ornate floor organ lamp
481	2- 10" carved Oriental figures of ladies...x2
482	2- 7 1/2" cloisonné vases...x2
483	8" carved jade covered urn
484	6" carved hardstone covered jar
485	Bronze & shell inkwell stand
486	12" Arts & Crafts style copper bowl
487	12 carved Oriental figures...x12
488	13 assorted Victorian, etc fans...x13
489	6 1/4" carved jade covered jar
490	Victorian copper & brass lamp (as seen)
491	2- 25" Oriental bronze crane figures (1 as seen)...x2
492	R J Horner birdseye maple bed, highboy with mirror & chest...lot
493	Dwarf oak bookcase
494	Carved Queen Anne style side chair
495	Victorian style upholstered chair
496	Handmade Oriental rug (approx 4'x 6')
497	Ornate cast iron fireplace surround
498	Handmade Ispahan carpet (approx 9 1/2'x 13')
499	2 ornate 11" silver-plate candlesticks...x2
500	4 piece Deco sterling dresser set...lot
501	8" Fuschia Roseville pottery vase (base repair)

THANKS FOR COMING!

Please join us for our upcoming auctions:

Antique, Collectibles & Estate

August 1st - Monday – 6 pm

Preview: Sunday 5 pm - 8 pm & Monday 10 am - 6 pm

Antique, Collectibles & Estate

August 8th - Monday – 6 pm

Preview: Sunday 5 pm - 8 pm & Monday 10 am - 6 pm

Checkerboard

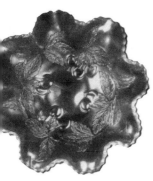

Cherry (Dugan)

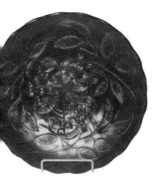

Cherry (Millersburg)

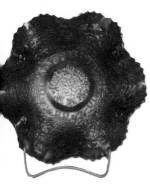

Cherry Chain

	M	A	G	B	PO	AO	Pas	Red
CHECKERBOARD BOUQUET								
Plate, 8"..............................		50						
CHECKERBOARD (WESTMORELAND)								
Cruet, Rare.........................							600*CL	
Pitcher, Rare.......................		3,800						
Tumbler, Rare	600	450						
Goblet, Rare......................	260	295						
Punch Cup	75	90						
Wine, Rare.........................	250							
Vase		2,000						
CHECKERS								
Ashtray	30							
Bowl 4"	18							
Bowl 9"	32							
Butter, 2 Sizes	140							
Plate, 7"	50							
CHERRY (DUGAN)								
Bowl, 5", Flat	30	40						
Bowl, 8" Flat......................	50	60			200			
Bowl, 8½", Ftd	165	275			390			
Plate, 6".............................		125			375			
Cruet, Rare.........................							500*W	
CHERRY (FENTON)								
(See Mikado Compote)								
CHERRY (MILLERSBURG)								
Bowl, 4".............................	30	50	55	495				
Bowl, 5", Hobnail								
Exterior Rare.......................				750				
Bowl, 7", Rare.....................	100	80	120					
Bowl, 9", Scarce..................	80	95	100					
Bowl, 10", Ice Cream	110	140	160	600				
Bowl, 9", Hobnail								
Exterior Rare.......................	700	985		1,800				
Compote, Large, Rare	900	1,000	1,100	2,800				
Banana Compote, rare.........		2,000		2,400			3,000V	
Milk Pitcher, Rare	620	625	625					
Plate, 6", Rare.....................	850							
Plate, 7½", Rare		800	950					
Plate, 10", Rare...................	4,000		4,350					
Powder Jar, Rare			1,250					
Covered Butter	180	250	295					
Covered Sugar	125	200	200					
Creamer or Spooner	125	190	190					
Pitcher, Scarce....................	950	750	850					
Tumbler, 2 Variations	250	450	250					
CHERRY BLOSSOMS								
Pitcher...............................				95				
Tumbler				28				
CHERRY AND CABLE (NORTHWOOD)								
Pitcher, Rare.......................	1,350							
Tumbler, Rare	400							
Bowl, 5", Scarce..................	45							
Butter, Rare........................	350							
Sugar, Creamer,								
Spooner, Each, Rare	175							
Bowl, 9", Scarce..................	95							
CHERRY AND CABLE INTAGLIO (NORTHWOOD)								
Bowl, 10"..........................	36							
Bowl, 5".............................	50							
CHERRY CHAIN (FENTON)								
Bon Bon	37	50	52	47			80W	1,650
Bowl, 7½"-10"	42	65	70	60			75W	2,000RS
Plate, 7"-9"	85			110			150W	
CHERRY CIRCLES (FENTON)								
Bon Bon	36	60		55				1,950
Bowl, 8".............................	40	60		56				
Compote............................	50	62		60			115W	
Plate, 9" Rare	450			120			175W	
CHERRY AND DAISIES (FENTON)								
Banana Boat	800			960				
CHERRY SMASH (U.S. GLASS)								
Bowl, 8".............................	50							
Butter	110							
Tumbler	135							
CHERRY STIPPLED								
Tumbler	75							
CHERUB								
Lamp, Rare.........................							425	

Christmas Compote

Chrysanthemum

Coin Spot

Colonial Lady

	M	A	G	B	PO	AO	Pas	Re(d)
CHIPPENDALE SOUVENIR								
Creamer or Sugar	65	80						
CHRISTMAS COMPOTE								
Large Compote, Rare	2,600	2,600						
CHRYSANTHEMUM (FENTON)								
Bowl, 9" Flat	65		85	95			90	1,080
Bowl, 10" Ftd	60		90	80			115	1,175
CHRYSANTHEMUM DRAPE								
Oil Lamp, Rare							900*	
CIRCLE SCROLL (DUGAN)								
Compote, Scarce		125						
Bowl, 10"	57	75						
Bowl, 5"	40	45						
Pitcher, Rare	1,850	2,600						
Tumbler, Rare	300	525						
Hat Shape, Rare	90	120						
Creamer or Spooner	150	200						
Vase Whimsey, Rare	120	140						
Butter or Sugar	250	350						
CLASSIC ARTS (CZECH)								
Vase, 7" (Egyptian)	445							
Powder Jar	475							
Rose Bowl	460							
Vase, 10", Rare	500							
CLEOPATRA								
Bottle	100							
CLEVELAND MEMORIAL (MILLERSBURG)								
Ashtray, Rare	2,450	2,000						
COAL BUCKET (U.S. GLASS)								
One Size	90		175					
COBBLESTONES (DUGAN)								
Bowl, 9"	55	70						
Bowl, 5"	38	40						
COBBLESTONES (DUGAN-IMPERIAL)								
Plate, Rare		1,000						
COBBLESTONES (IMPERIAL)								
Bowl, 5"	30	40	45					
Bowl, 8½"	50	70	75					
Bon Bon	40	55	55				75AM	
COIN DOT (FENTON)								
Bowl, 6"-10"	36	50	40			210	85LV	1,100
Plate, 9", Rare	150	170	180	160				
Pitcher, Rare	295	425	450	410				
Tumbler, Rare	75	125	125	100				
Basket Whimsey, Rare	60			85				
Rose Bowl	55		75	70				1,100
COIN DOT VT. (WESTMORELAND)								
Compote	60				170MO	225	275IBO	
Rose Bowl	65	75					290TL	
Bowl	40	55	65		180	195	75	
COIN SPOT (DUGAN)								
Compote	40	58	62	70	190	300		
Goblet, Rare							250IG	
COLOGNE BOTTLE (CAMBRIDGE)								
One Size, Rare	600		850					
COLONIAL (IMPERIAL)								
Lemonade Goblet	60							
Vase	38	58						
Toothpick Holder	50	90	85					
Open Sugar or Creamer	36		45					
Candlesticks, Pair	185							460
COLONIAL LADY (IMPERIAL)								
Vase, Rare	900	1,025						650
COLUMBIA (IMPERIAL)								
Vase	38	48	45					
Compote	45	60	57					
COLUMBUS								
Plate, 8"	38							
COMPASS (DUGAN)								
(Exterior Pattern Only)								
COMPOTE VASE								
Stemmed Shape	50	55	58	60				
CONCAVE DIAMONDS (DUGAN)								
Pitcher w/lid			450RG				550V	
Tumbler			400RG				190V	
Pickle Castor, Ornate Holder	450*							
Tumble-Up, Complete, Rare			375RG			600*	350IB	
Coaster, Not Iridized							20	

Constellation

Coral

Cosmos and Cane

Country Kitchen

	M	A	G	B	PO	AO	Pas	Red
CONCAVE FLUTE (WESTMORELAND)								
Rose Bowl	50		65					
Vase	40	60	65					
CONCORD (FENTON)								
Bowl, 9", Scarce	160	185	200	195			195AM	
Plate, 10", Rare	400	700	625				600AM	
CONE AND TIE (IMPERIAL)								
Tumbler, Rare		650						
CONNIE (NORTHWOOD)								
Pitcher							650W	
Tumbler							90W	
CONSTELLATION (DUGAN)								
Compote	40	55			160		85V	
CONTINENTAL BOTTLE								
2 Sizes	35							
COOLEEMEE, NC (FENTON)								
Advertising Plate, Rare (Heart and Vine)	1,000							
CORAL (FENTON)								
Bowl, 9"	85		115	95			125V	
Compote, Rare	265		295				325W	
Plate, 9½", Rare	750		850	850				
CORINTH (DUGAN)								
Bowl, 9"	38	50			170			
Banana Dish	50	75			280		150AP	
CORINTH (WESTMORELAND)								
Vase	28	45		50TL			60SM	
Bowl	38	50		60TL				
CORN BOTTLE (IMPERIAL)								
One Size	285		265				275AM	
CORN CRUET								
One Size, Rare							900W	
CORN VASE (NORTHWOOD)								
Regular Mold	675	585	500	1,500		2,700	2,300IB	
Pulled Husk, Rare		6,000	6,000					
Fancy Husk, Rare (Dugan)	800*						900*W	
CORNING (CORNING)								
Insulator	20+							
CORNUCOPIA (FENTON)								
Candlestick, 5", pr	75						140W	
Vase, 5"	65						100W	
Candle Holder, 6½"							100W	
CORONATION (ENGLISH)								
Vase, 5" (Victoria Crown Design)	75*							
COSMOS (MILLERSBURG)								
Bowl, 5"			55					
Plate, 6"			65					
COSMOS AND CANE								
Bowl, 10"	58						110W	
Bowl, 5"	36						70W	
Compote, Tall, Rare	300						285W	
Butter, Covered	175						295W	
Covered Sugar or Creamer	125						225W	
Flat Tray, Rare							150	
Spooner	95						195W	
Stemmed Dessert							140W	
Pitcher, Rare	665						1,250W	
Tumbler, Rare	85						150W	
Advertising Tumbler, Rare	235							
Rose Bowl	95	150					165AM	
Rose Bowl Whimsey	700							
Spittoon Whimsey, Rare	4,000						3200W	
Chop Plate, Rare	385						475W	
COSMOS VT. (FENTON)								
Bowl, 9"-10"	35	40		80			60V	425
Plate, 10", Rare	85	110			300			
COUNTRY KITCHEN (MILLERSBURG)								
Bowl, 9", Rare	250							
Bowl, 5", Rare	100							
Spittoon Whimsey, Rare		3,800						
Covered Butter, Rare	500	600						
Sugar, Creamer or Spooner	375	325	550*				800V	
Vase Whimsey, Rare	500	550						
COURTHOUSE (MILLERSBURG								
Lettered (Round or Ruffled), Scarce		750						
Unlettered, Rare		1,250						

Crackle

Crystal Cut

Cut Arches

Dahlia

	M	A	G	B	PO	AO	Pas	Rec
COVERED FROG (HEISEY)								
One Size		300					400	
COVERED HEN (ENGLISH)								
One Size	90			110				
COVERED LITTLE HEN (TINY)								
Miniature, 3½", Rare							90CM	
COVERED MALLARD (U.S. GLASS)								
One Shape							450CM	
COVERED SWAN (ENGLISH)								
One Size	150	240						
COVERED TURKEY (HEISEY)								
One Size		385						
COVERED TURTLE (HEISEY)								
One Size			300				400PK	
CR (ARGENTINA)								
Ash Tray	75			85				
CRAB CLAW (IMPERIAL)								
Bowl, 5"	25	35	35					
Bowl, 10"	45	58	60				40SM	
Fruit Bowl w/base	90							
Pitcher, Scarce	395							
Tumbler, Scarce	95							
Cruet, Rare	850							
CRACKLE (IMPERIAL)								
Auto Vase	25	35	32					
Bowl, 9"	18	30	27					
Bowl, 5"	12	18	15					
Candy Jar w/lid	26		35					
Candlestick, 3½"	25							
Candlestick, 7"	30							
Plate	30	45	45					
Punch Bowl and Base	48		60					
Punch Cup	10		16					
Pitcher, Dome Base	85	140	140					
Spittoon, Large	38							
Tumbler, Dome Base	18	26	26					
Wall Vase	35							
Window Planter, Rare	95							
CROCEUS VT.								
Tumbler	38							
CRUCIFIX (IMPERIAL)								
Candlestick, Rare, Ea	600							
CRYSTAL CUT (CRYSTAL)								
Compote	50							
CUT ARCHES (ENGLISH)								
Banana Bowl	50							
CUT ARCS (FENTON)								
Bowl, 7½"-10"	32							
Compote	40	48		45				
Vase Whimsey (From Bowl)	38	45		42			70W	
CUT COSMOS (MILLERSBURG)								
Tumbler, Rare	395							
CUT CRYSTAL (U.S. GLASS)								
Compote, 5½"	90							
Water Bottle	165							
CUT FLOWERS (JENKINS)								
Vase, 10"	150						195SM	
CUT OVALS (FENTON)								
Candlesticks, pr	165						190	600
Bowl, 7"-10"	50						70	385
CUT SPRAYS								
Vase, 9"	35				95			
DAHLIA (DUGAN)								
Bowl, 10", Ftd	90	130					275W	
Bowl, 5", Ftd	40	50					110W	
Butter	110	150					325W	
Sugar	80	100					250W	
Creamer or Spooner	75	100					250W	
Pitcher, Rare	400	850					700W	
Tumbler, Rare	75	135					150W	
DAHLIA (FENTON)								
Twist Epergne, One Lily	250						300W	

238

	M	A	G	B	PO	AO	Pas	Red
DAHLIA AND DRAPE (FENTON)								
Tumble-Up, Complete	150						185IB	
DAINTY BUD VASE								
One Size	50							
DAISY (FENTON)								
Bon Bon, Scarce	185		195					
DAISY BASKET (IMPERIAL)								
One Size	65						80SM	
DAISY BLOCK (ENGLISH)								
Rowboat, Scarce	265	290					295AQ	
DAISY AND CANE (ENGLISH)								
Decanter, Rare	75							
Spittoon, Rare				185				
DAISY CHAIN								
Shade	45							
DAISY CUT BELL (FENTON)								
One Size, Rare	475							
DAISY AND DRAPE (NORTHWOOD)								
Vase	290	345	1,950	495		660	950IB	
DAISY IN OVAL PANELS (U.S. GLASS)								
Creamer or Sugar, Ea	50							
DAISY AND PLUME (NORTHWOOD)								
Rose Bowl, 2 Shapes	55	65	70	75	125	1,800	500	
Compote	50	60	65	70	110			
Candy Dish	45	55	60	65	100			
DAISY SQUARES								
Rose Bowl	275	400	500				250AM	
Goblet, Rare							350AM	
Compote, Rare							500AM	
DAISY WEB (DUGAN)								
Hat, Rare	80	95			125			
DAISY WREATH (WESTMORELAND)								
Bowl, 8"-10"						490BO	440MO	
DANCE OF THE VEILS (FENTON)								
Vase, Rare	2,650							
DANDELION (NORTHWOOD)								
Mug	400	410	565	450		485	650BO	
Pitcher	375	575	650	790			2,000W	
Tumbler	50	50	65	90			190W	
Vase Whimsey, Rare		650						
DECO LILY								
Bulbous Vase	85							
DEEP GRAPE (MILLERSBURG)								
Compote, Rare	1,000	1,200	1,300	2,400				
Compote, Ruffled Top, Rare		1,850						
Rose Bowl, Stemmed			1,950*					
DeVILBISS								
Atomizer, Complete	40						45	
Perfumer	40						45	
DIAMONDS (MILLERSBURG)								
Pitcher	235	285	265				225AO	
Tumbler	60	80	75				90AO	
Punch Bowl and Base, Rare	2,000	1,800	1,800					
Pitcher Oddity (No Spout)		350	350					
DIAMOND BAND (CRYSTAL)								
Open Sugar	38	45						
Float Set	250	300						
DIAMOND BAND AND FAN (ENGLISH)								
Cordial Set, Complete, Rare	750*							
DIAMOND CHECKERBOARD								
Cracker Jar	75							
Butter	70							
Bowl, 9"	35							
Bowl, 5"	25							
Tumbler	85							
DIAMOND DAISY								
Plate, 8"	75							

Daisy and Drape

Daisy Squares

Daisy Wreath

Deep Grape

239

Diamond Fountain

Diamond Lace

Diamond and Rib

Dolphins

	M	A	G	B	PO	AO	Pas	Red
DIAMOND AND DAISY CUT (U.S. GLASS)								
Vase, 10", Square	85							
Compote	45	60		60				
Pitcher, Rare	325			350				
Tumbler, Rare	45			50				
DIAMOND AND DAISY CUT VT (JENKINS)								
Punch Bowl/Base	500							
DIAMOND AND FILE								
Banana Bowl	55							
Bowl, 7"-9"	35						58	
DIAMOND FLUTES (U.S. GLASS)								
Creamer	35							
Parfait	40							
DIAMOND FOUNTAIN (HIGBEE)								
Cruet, Rare	675*							
DIAMOND LACE (IMPERIAL)								
Bowl, 10"-11"	45	65						
Bowl, 5"	28	37						
Pitcher		285						
Tumbler	75	66					200W*	
DIAMOND OVALS (ENGLISH)								
Compote (Open Sugar)	35							
Creamer	35							
DIAMOND PINWHEEL (ENGLISH)								
Compote	45							
Butter	65							
DIAMOND POINT								
Basket, Rare	785	850		850				
DIAMOND POINT COLUMNS (IMPERIAL)								
Bowl, 4½"	20							
Compote	26						35	
Plate, 7"	35							
Vase	30	38	40				40	
Butter	65							
Creamer, Sugar or Spooner, ea	40							
Powder Jar w/lid	50							
DIAMOND POINTS (NORTHWOOD)								
Vase, 7"-14"	30	40	37	42	275*	490	280IG	
DIAMOND PRISMS (ENGLISH)								
Compote	45							
DIAMOND AND RIB (FENTON)								
Vase, 7"-12"	25	35	32	32			40SM	
Funeral Vase, 17"-22"	575	850	965	975			700W	
Vase Whimsey			600					
DIAMOND RING (IMPERIAL)								
Bowl, 9"	37	45					40	
Bowl, 5"	20	28					26	
Fruit Bowl, 9½"	52	85					57	
DIAMOND STAR								
Mug, 2 sizes	85							
Vase, 8"	60							
DIAMOND AND SUNBURST (IMPERIAL)								
Bowl, 8"	48	50	48				55AM	
Decanter	100	140	150					
Wine	40	48	52					
Oil Cruet, Rare		800*						
DIAMOND TOP (ENGLISH)								
Creamer	32							
Spooner	32							
DIAMOND VANE (ENGLISH)								
Creamer, 4"	35							
DIVING DOLPHINS (ENGLISH)								
Bowl, 7", Ftd	210	250	275	260				
DOG								
Ashtray	65							
DOGWOOD SPRAYS (DUGAN)								
Compote	40				160			
Bowl, 9"	37	50			165			
DOLPHINS (MILLERSBURG)								
Compote Rare		1,900	1,975	4,200				

	M	A	G	B	PO	AO	Pas	Red

Double Fan Tumbler

Double Star

Dragon and Lotus

Drapery

DORSEY AND FUNKENSTEIN (NORTHWOOD)

	M	A	G	B	PO	AO	Pas	Red
Plate		335						

DOTS AND CURVES

	M	A	G	B	PO	AO	Pas	Red
Hatpin		42						

DOTTED DAISIES

| Plate, 8" | 65 | | | | | | | |

DOUBLE DOLPHINS (FENTON)

	M	A	G	B	PO	AO	Pas	Red
Bowl, 9"-11", Ftd							110	
Cake Plate, Center Handle							70	
Candlesticks, pr							85	
Compote							60	
Fan Vapse							65	
Covered Candy Dish, Stemmed							75	
Bowl, 8"-10", Flat							60	

DOUBLE DUTCH (IMPERIAL)

	M	A	G	B	PO	AO	Pas	Red
Bowl, 9", Ftd	45	60	60				65SM	

DOUBLE FAN (ENGLISH)

| Tumbler, Rare | 100 | | | | | | | |

DOUBLE LOOP (NORTHWOOD)

	M	A	G	B	PO	AO	Pas	Red
Creamer	55	65	70	100		260		
Sugar	55	60	70	100		265		

DOUBLE SCROLL (IMPERIAL)

	M	A	G	B	PO	AO	Pas	Red
Bowl	40	52	50				65	185
Candlesticks pr	50	60	60				70	200
Punch Cup	22							

DOUBLE STAR (CAMBRIDGE)

	M	A	G	B	PO	AO	Pas	Red
Pitcher, Scarce	600	595	495					
Tumbler, Scarce	160	110	50					
Spittoon Whimsey, Rare			2,500					
Bowl, 9", Rare			295*					

DOUBLE STEM ROSE (DUGAN)

	M	A	G	B	PO	AO	Pas	Red
Bowl, 8½", Dome Base	38	45	50	55	225		60LV	

DOUGHNUT BRIDLE ROSETTE

	M	A	G	B	PO	AO	Pas	Red
One Size		75						

DRAGON AND LOTUS (FENTON)

	M	A	G	B	PO	AO	Pas	Red
Bowl, 9", Flat	58	80		75	750	860	800IM	1,000
Bowl, 9", Ftd	55	67	85	70	680		130	1,150
Plate, 9½", Rare	650	600		900	1,100		800	3,000

DRAGON AND STRAWBERRY (FENTON)

	M	A	G	B	PO	AO	Pas	Red
Bowl, 9", Flat, Scarce	350		425	475				
Bowl, 9", Ftd Scarce	365		450	490				
Plate (Absentee Dragon), Rare	2,500							

DRAGONFLY

| Shade | | | | | | | 48 | |

DRAGONFLY LAMP

| Oil Lamp, Rare | | | | | | | 1,650 | |

DRAGON'S TONGUE (FENTON)

	M	A	G	B	PO	AO	Pas	Red
Bowl, 11" Scarce	385							
Shade					95			

DRAPE AND TASSEL

| Shade | 38 | | | | | | 40W | |

DRAPERY (NORTHWOOD)

	M	A	G	B	PO	AO	Pas	Red
Vase	30	40	42	48			54	
Rose Bowl	60	75	85	85		500	275	
Candy Dish	55	67	75			460	195IB	

DRAPERY VT. (FENTON)

	M	A	G	B	PO	AO	Pas	Red
Pitcher, Rare	485							
Tumbler, Scarce	70							

DREIBUS PARFAIT SWEETS (NORTHWOOD)

	M	A	G	B	PO	AO	Pas	Red
Plate, 6", Handgrip		375						

DUCKIE

| Powder Jar w/lid | 35 | | | | | | | |

241

Elks (Millersburg)

Elks (Fenton)

Emu

Estate

	M	A	G	B	PO	AO	Pas	Re
DUGAN FAN (DUGAN)								
Sauce, 5"	38	48						
Gravy Boat, Ftd	50	60			145		90	
DUNCAN (NATIONAL GLASS)								
Cruet	410*							
DURAND (FENTON)								
Bowl - Grape and Cable				900				
DUTCH MILL								
Plate, 8"	35							
Ashtray	35							
DUTCH PLATE								
One Size, 8"	40							
DUTCH TWINS								
Ashtray	45							
EAGLE FURNITURE (NORTHWOOD)								
Plate		360						
EBON								
Vase		90BA						
ELEGANCE								
Bowl, 8¼"		190						
Plate							2,500	
ELKS (DUGAN)								
Nappy, Very Rare		3,500						
ELKS (FENTON)								
Detroit Bowl, Scarce	1,000*		350	325				
Parkersburg Plate, Rare			950	800				
Atlantic City Plate, Rare			1,150	975				
Atlantic City Bowl				395				
1911 Atlantic City Bell, Rare				850				
1914 Parkersburg Bell, Rare				850				
ELKS (MILLERSBURG)								
Bowl, Rare		1,250						
Paperweight, Rare		1,200	1,350					
EMBROIDERED MUMS (NORTHWOOD)								
Bowl, 9"	300	250	260	375		1,175	95	
Stemmed Bon Bon							115	
Plate	145	170		175			875IG	
EMU (CRYSTAL)								
Bowl, 5", Rare	60							
Bowl, 10", Rare	180						450AM	
ENAMELLED GRAPE (NORTHWOOD)								
Pitcher				340				
Tumbler				40				
ENGLISH BUTTON BAND (ENGLISH)								
Creamer	38							
Sugar	38							
ENGLISH HOB AND BUTTON (ENGLISH)								
Bowl, 7"-10"	60	75	90	60				
Epergne (metal base), Rare	100			110				
ENGRAVED FLORAL (FENTON)								
Tumbler			85					
ENGRAVED GRAPES (FENTON)								
Vase, 8"	40						50	
Candy Jar w/lid	45							
Juice Glass	20							
Pitcher, Squat	80							
Tumbler	20							
Pitcher, Tall	85							
Tumble-Up	135							
ESTATE (WESTMORELAND)								
Mug, Rare	85						95	
Perfume							110AQ	
Creamer or Sugar	48					120	85	
Bud Vase, 6"	40						75	
ESTATE, STIPPLED (WESTMORELAND)								
Vase, 3"					85		95	
EXCHANGE BANK (NORTHWOOD)								
Plate, 6"		200						
EYE CUP								
One Size	55							

	M	A	G	B	PO	AO	Pas	Red
FAMOUS								
Puff box	75							
FANCIFUL (DUGAN)								
Bowl, 8½"	45	58			275		60	
Plate, 9"	225	250		275	410		260	
FANCY (NORTHWOOD)								
(Interior on some "Fine Cut and Roses" Rose Bowls)								
FANCY CUT (ENGLISH)								
Miniature Pitcher, Rare	125							
Miniature Tumbler	55							
FANCY FLOWERS (IMPERIAL)								
Compote	95		110					
FANS (ENGLISH)								
Pitcher	60							
Cracker Jar (metal lid)	118							
FANTAIL (FENTON)								
Bowl, 5", Ftd	75			180			90	
Bowl, 9", Ftd	60			135			42	
Compote	70			160				
Plate, Ftd, Rare				1,100*				
FARMYARD (DUGAN)								
Bowl, 10", Rare		2,800	7,000		9,600			
Plate, 10½"		8,500						
FASHION (IMPERIAL)								
Creamer or Sugar	42	110					48	
Fruit Bowl and Base	60		85				65	
Punch Bowl and Base	70	165					85	5,000*
Punch Cup	22	35					26	400*
Pitcher	170	900					500SM	
Tumbler	20	200					90SM	
Bowl, 9"	38		75				47SM	
Bride's Basket	100						120	
Butter	65	185						
Rose Bowl Whimsey, Rare	300	1,150	700					
FEATHER AND HEART (MILLERSBURG)								
Pitcher, Scarce	585	700	775					
Tumbler, Scarce	58	97	105					
FEATHER STITCH (FENTON)								
Bowl, 8½"-10"	50	75	68	60				
FEATHER SWIRL (U.S. GLASS)								
Vase	50							
Butter	110							
FEATHERED ARROW (ENGLISH)								
Bowl, 8½"	40							
FEATHERED SERPENT (FENTON)								
Bowl, 5"	26	45	40	37				
Bowl, 10"	48	70	65	60				
Spittoon Whimsey, Rare		3,500						
FEATHERS (NORTHWOOD)								
Vase, 7"-12"	28	40	45					
FELDMAN BROTHERS (NORTHWOOD)								
Bowl		250						
FENTONIA								
Bowl, 9½", Ftd	60	85	75	70				
Bowl, 5", Ftd	27	48	42	38				
Fruitbowl, 10"	75			90				
Butter	110			170				
Creamer, Sugar or Spooner	70			85				
Pitcher	385			560				
Tumbler	45			70				
FENTONIA FRUIT (FENTON)								
Bowl, 6", Ftd	45			55				
Bowl, 10" Ftd	100			150				
Pitcher, Rare	475			670				
Tumbler, Rare	120			150				
FENTON'S ARCHED FLUTE (FENTON)								
Toothpick Holder	75			95			130	

Fancy Cut

Fashion

Feather and Heart

Feather Stitch

Fern Panels

Field Flower

Field Thistle

Fine Rib

	M	A	G	B	PO	AO	Pas	R
FENTON'S BASKET (FENTON)								
Two Row or Three Row								
(Open Edge)	35		40	38			685IB	4.
Advertising	45							
FERN (FENTON)								
Bowl, 7"-9", Rare				800*				
FERN (NORTHWOOD)								
Bowl, 6½"-9"	46	50	50					
Compote	45	50	50				80	
Hat, Rare	52	72	72				90	
FERN BRAND CHOCOLATES (NORTHWOOD)								
Plate		345						
FERN PANELS (FENTON)								
Hat	35		46	40			495	
FIELD FLOWER (IMPERIAL)								
Pitcher, Scarce	150	350	365	390			350AM	
Tumbler, Scarce	35	60	70	150			90AM	
Milk Pitcher, Rare	165	170	185				210CM	
FIELD THISTLE (U.S. GLASS)								
Plate, 6", Rare	180							
Plate, 9" Rare	350							
Butter, Rare	125							
Sugar, Creamer or								
Spooner, Rare	70							
Compote, Large	85							
Pitcher, Scarce	150							
Tumbler, Scarce	38							
Breakfast Set, 2 Pc, Rare							200IB	
Bowl, 6"-10"	45						250LG	
Vase	60							
FILE (IMPERIAL AND ENGLISH)								
Pitcher, Rare	265	435						
Tumbler, Scarce	158							
Bowl, 5"	30	40						
Bowl, 7"-10"	40	50					60	
Compote	40	45					50	
Vase	80							
Butter	195							
Creamer or Spooner	100							
Sugar	120							
FILE AND FAN								
Bowl, 6", Ftd	38				156			
Compote						250	125MO	
FINE BLOCK (IMPERIAL)								
Shade			40					
FINE CUT FLOWER AND VT. (FENTON)								
Compote	50							
Goblet	50							
FINE CUT HEART (MILLERSBURG)								
(Primrose Bowl Exterior Pattern)								
FINE CUT OVALS (MILLERSBURG)								
(Whirling Leaves Exterior Pattern)								
FINE CUT RINGS (ENGLISH)								
Oval Bowl	40							
Vase	50							
Celery	58							
Butter	70							
Creamer	45							
Stemmed Sugar	45							
Stemmed Cake Stand	70							
Round Bowl	35							
Jam Jar w/lid	50							
FINE CUT AND ROSES (NORTHWOOD)								
Rose Bowl, Ftd	50	66	75			975	450IB	
Candy Dish, Ftd	46	60	80			860	265W	
(Add 25% for Fancy Interior)								
FINE PRISMS AND DIAMONDS (ENGLISH)								
Vase, 7"-14"							80	
FINE RIB (NORTHWOOD, FENTON AND DUGAN)								
Bowl, 9"-10"	50	60	75					
Bowl, 5"	26	30	36					
Plate, 9"	70	90						
Vase, 7"-15", 2 Types	38	46	58	42			260VO	40(
Compote					150			

Fleur De Lis (Jenkins)

Floral and Optic

Flowering Dill

Flowering Vine

	M	A	G	B	PO	AO	Pas	Red
FISH NET (DUGAN)								
Epergne		585			525			
FISH VASE (CZECH)								
One shape, Marked "Jain"	75	95	90					
FISHERMAN'S MUG (DUGAN)								
One Size	335	150		450	1,350		225HO	
FISHSCALE AND BEADS (DUGAN)								
Bowl, 6"-8"	26	34			150		50	
Bride's Basket, Complete					120			
Plate, 7"	40							
FIVE HEARTS (DUGAN)								
Bowl, 8¼", Dome Base	48	52			165			
FIVE LILY EPERGNE								
Complete, Metal Fittings	175	250						
FLANNEL FLOWER (CRYSTAL)								
Compote, Large	90	140						
Cake Stand	120	175						
FLARED PANEL								
Shade					50MO			
FLARED WIDE PANEL								
Atomizer, 3½"	90							
FLEUR-DE-LIS (JENKINS)								
Vase	195							
FLEUR-DE-LIS (MILLERSBURG)								
Bowl, 8½", Flat	210	290	275			240CM		
Bowl, 8½", Ftd	350	475	450					
Rose Bowl, Either Base, Rare		4,300						
FLICKERING FLAMES								
Shade							45	
FLORA (ENGLISH)								
Float Bowl				75				
FLORAL								
Hatpin							50AM	
FLORAL FAN								
Etched Vase	42							
FLORAL AND GRAPE (DUGAN)								
Pitcher	195	285		270			420W	
Tumbler	28	35		40			48W	
Hat	35							
FLORAL AND GRAPE VT. (FENTON)								
Pitcher, 2 Variations	195	285	290	270				
Tumbler	28	35	38	30				
FLORAL AND OPTIC (IMPERIAL)								
Bowl, 8"-10", Ftd	28				168		38	420
Bowl, 8"-10", Flat	30						20	
Cake Plate, Ftd					180		50	600
Rose Bowl, Ftd					190		180AQ	
FLORAL OVAL (HIG-BEE)								
Bowl, 8"	45							
Plate, 7", Rare	80							
Creamer	65							
FLORAL AND SCROLL								
Shade, Various Shapes	36+							
FLORAL AND WHEAT (U.S. GLASS)								
Compote	38	45		42	158			
Bon Bon, Stemmed	38	45		42	155			
FLORENTINE (FENTON AND NORTHWOOD)								
Candlesticks, pr	80			115			195	
FLORENTINE (IMPERIAL)								
Hat Vase							85	
FLOWER BASKET								
One Size	48							
FLOWER AND BEADS								
Plate, 7½", 6 Sided	95	110						
Plate, 8½", Round	75							
FLOWER POT (FENTON)								
One Size, Complete	45						65	
FLOWERING DILL (FENTON)								
Hat	36		42	40				500
FLOWERING VINE (MILLERSBURG)								
Compote, Tall, Very Rare		3,800	3,800					

245

Flowers and Frames

Fluted Scroll

Footed Prism Panels

Forks

	M	A	G	B	PO	AO	Pas	Re
FLOWERS AND FRAMES (DUGAN)								
Bowl, 8"-10"	50	65	68		300		95	
FLOWERS AND SPADES (DUGAN)								
Bowl, 10"	46	80	84		195			
Bowl, 5"	22	30	35		136			
FLUTE (BRITISH)								
Sherbet, Marked "British"	45							
FLUTE (MILLERSBURG)								
Vase, Rare	275	300	350					
Bowl, 4" (Variant)		40						
Compote, 6" (Clover Base) Marked "Crystal,"								
Very Rare	450	500						
Punch Bowl and Base, Rare.	250	325						
Punch Cup	25	30						
Bowl, 10"	65	85						
Bowl, 5"	20	30						
FLUTE (NORTHWOOD)								
Creamer or Sugar	65	85	90					
Salt Dip, Ftd	30							
Sherbet	35	47	45				75V	
Pitcher, Rare	385		595				50	
Tumbler, 3 Varieties	48		70					
Ringtree, Rare	175							
Bowl, 10"	45	50						
Bowl, 5"	22	26						
Butter	120	165	175					
FLUTE #3 (IMPERIAL)								
Covered Butter	175	195	190					
Sugar, Creamer or Spooner	85	95	90					
Celery, Rare		285*						
Punch Bowl and Base	275	475	495					
Punch Cup	25	35	38					
Pitcher	320	580	495	495				
Tumbler	45	180	185	85				
Handled Toothpick Holder	60						80SM	
Toothpick Holder, Regular	55	70	75	90			295AQ	
Bowl, 10"		220						
Bowl, 5"	30	65						
Custard Bowl, 11"		285	300					
Cruet	85							
FLUTE AND CANE (IMPERIAL)								
Pitcher, Stemmed, Rare	385						900W	
Wine	40							
Champagne, Rare	120							
Milk Pitcher	120							
Punch Cup	28							
Tumbler, Rare	350*							
FLUTED SCROLL (DUGAN)								
Rose Bowl, Ftd, Very Rare		950						
FLYING BAT								
Hatpin, Scarce	35	55	58				70	
FOLDING FAN (DUGAN)								
Compote		65		75	120	285		
FOOTED DRAPE (WESTMORELAND)								
Vase	40						50W	
FOOTED PRISM PANELS (ENGLISH)								
Vase	60		75	85				
FOOTED RIB (NORTHWOOD)								
Vase	40	65	80	70			100	
Vase, Advertising						190	140	
						235		
FOOTED SHELL (WESTMORELAND)								
Large, 5"	30	40	45	50	100MO		60AM	
Small, 3"	35	45	50	54			70AM	
FORGET-ME-NOT (FENTON)								
Pitcher	165	345	375	385			320W	
Tumbler	26	45	48	38			50W	
FORKS (CAMBRIDGE)								
Cracker Jar, Rare			495*					
FORMAL (DUGAN)								
Hatpin Holder, Rare	175	160						
Vase, Jack-In-Pulpit, Rare	120	100					150	
49'ER (IMPERIAL)								
Tumbler	75							
Wine	50							
Decanter	125							
Pitcher, Squat	250							

474 (Imperial)

Frosted Block

Fruit Basket

Fruits and Flowers

	M	A	G	B	PO	AO	Pas	Red
FOSTORIA #600 (FOSTORIA)								
Napkin Ring	75							
FOSTORIA #1231 (FOSTORIA)								
Rose Bowl...........................							100	
FOSTORIAL #1299 (FOSTORIA)								
Tumbler	75*							
FOUNTAIN LAMP								
Complete, Scarce.................	265							
FOUR FLOWERS								
Plate, 6½"		70	75		190			
Plate, 9"-10½"		700			475			
Bowl, 6¼"		38	45	65	160			
Bowl, 10"		65	70	90	185			
FOUR FLOWERS VT.								
Bowl, 9"-11"	45	60	65		170		90TL	
Bowl, 8½", Ftd		70	75				100TL	
Plate, 10½", Rare		650	650		400		400SM	
Bowl on Metal Base, Rare					200		350TL	
474 (IMPERIAL)								
Bowl, 8"-9"	60		85					
Punch Bowl and Base.........	250	975	600					
Cup....................................	28	40	35					
Covered Butter	95	175	125					
Creamer, Sugar or								
Spooner......................	60	110	85					
Milk Pitcher, Scarce............	225		475					
Pitcher, 2 Sizes, Scarce........	180	585	495				600PK*	
Tumbler, Scarce	35	80	65				120PK*	
Goblet	50	90	65					
Wine, Rare.........................	75							
Cordial, Rare	90	150						
Vase 7", Rare								2,500
Vase, 14", Rare	950		1,200					
FRENCH KNOTS (FENTON)								
Hat....................................	36	40	50	42				
FROLICKING BEARS								
(U.S. GLASS)								
Pitcher, Rare.......................			9,200					
Tumbler, Rare			8,000					
FROSTED BLOCK								
(IMPERIAL)								
Bowl, 6½"-7½".....................	26						32CM	
Bowl, 9"...............................	32						36CM	
Celery Tray.........................	40							
Covered Butter	67							
Creamer or Sugar	50							
Rose Bowl...........................	45						75CM	
Compote.............................	85						120CM	
Milk Pitcher, Rare...............	90							
Pickle Dish, Handled, Rare ..	40						58CM	
Bowl, Square, Rare..............	40						52CM	
Plate, 7½"............................							60CM	
Plate, 9"..............................							65CM	
25% More if Marked								
"Made in USA"								
FROSTED BUTTONS (FENTON)								
Bowl, 10", Ftd.....................							175	
FROSTED RIBBON								
Pitcher...............................	80							
Tumbler	27							
FROSTY								
Bottle	25							
FRUIT BASKET								
(MILLERSBURG)								
Compote, Handled, Rare......		1,500						
FRUIT AND BERRIES								
(ENGLISH)								
Bean Pot, Covered, Rare	250		275					
FRUIT AND FLOWERS								
(NORTHWOOD)								
Bowl, 9"...............................	58	70	70				85	
Bowl, 5"...............................	38	42	42				60	
Fruit Bowl, 10"	60	80					80	
Banana Plate, 7", Rare.........		225	210					
Bon Bon, Stemmed..............	38	50	50	65		540	575W	
Plate, 7".............................	185	200	210	240			230	
Plate, 9½"............................	200	225	225				350	
FRUIT JAR (BALL)								
One Size	65							
FRUIT LUSTRE								
Tumbler	38							

Garden Path

Garland

Gay 90's

Goddess of Harvest

	M	A	G	B	PO	AO	Pas	Red
FRUIT SALAD (WESTMORELAND)								
Punch Bowl, and Base, Rare	750	885			3,850			
Cup, Rare	30	37			40			
GAMBIER, MT. (CRYSTAL)								
Mug	45							
GARDEN MUMS (NORTHWOOD)								
Bowl, 8½"-10"	60	75	80	85			100	
Plate, 7", Regular or Handgrip	170	200	210	225			240	
GARDEN PATH (DUGAN)								
Bowl, 8½"-10"	40	70					85W	
Compote, Rare	175	300					450	
Fruit Bowl, 10"	85	95						
Plate, 6", Rare	425	600			1,050		510W	
Bowl, 5"					47			
GARDEN PATH VT. (DUGAN)								
Bowl, 9"					165			
Fruit Bowl, 10"		360			395			
Plate, 11", Rare		3,000			3,500			
GARLAND (FENTON)								
Rose Bowl, Ftd	45	70		65				
GAY 90'S (MILLERSBURG)								
Pitcher, Rare		8,500	9,500					
Tumbler, Rare	1,250	1,150						
GEORGIA BELLE (DUGAN)								
Compote, Ftd	65	75	80		170			
Card Tray, Ftd, Rare	65	85	90		180			
GOD AND HOME (DUGAN)								
Pitcher, Rare				1,000				
Tumbler, Rare				285				
GODDESS ATHENA								
Epergne, Rare			1,800				2,800AM	
GODDESS OF HARVEST (FENTON)								
Bowl, 9½", Rare	4,300	4,000		4,100				
Plate, Very Rare		6,700*						
GOLD FISH								
Bowl	70							
GOLDEN CUPIDS (CRYSTAL)								
Bowl, 9", Rare							465	
Bowl, 5", Rare							195	
GOLDEN FLOWERS								
Vase, 7½"	48							
GOLDEN GRAPES (DUGAN)								
Bowl, 7"	36	45	45				48	
Rose Bowl, Collar Base	50							
GOLDEN HARVEST (U.S. GLASS)								
Decanter, w/stopper	137	230						
Wine	28	35						
GOLDEN HONEYCOMB (IMPERIAL)								
Bowl, 5"	25							
Plate, 7"	45							
Compote	47							
Creamer or Sugar	32							
Bon Bon	40	50	50				60AM	
GOLDEN OXEN								
Mug	48							
GOLDEN WEDDING								
Bottle, Various Sizes	16+							
GOOD LUCK (NORTHWOOD)								
Bowl, 8¼"	215	235	265	275		850	900LV	
Plate, 9"	310	385	600	420			950IB	
GOOD LUCK VT. (NORTHWOOD)								
Bowl, 8¼", Rare	225	275	275				325	
GOODYEAR								
Ashtray in Tire	50							
GOOSEBERRY SPRAY								
Bowl, 10"	50	85	90	90			115	
Bowl, 5"		110	125	125			140	
Compote, Rare		210	225	225			240	
GOTHIC ARCHES								
Vase, 8"-12", Rare	40	60	70				70SM	
GRACEFUL (NORTHWOOD)								
Vase	50	70	75	80			185W	
GRAND THISTLE (FINLAND)								
Pitcher, Rare				1,800				
Tumbler, Rare				550				

Grape and Cable (Fenton)

Grape (Imperial)

Grape and Cable (Northwood)

	M	A	G	B	PO	AO	Pas	Red
GRAPE (FENTON'S GRAPE AND CABLE)								
Orange Bowl, Ftd................	85	185	195	175			200	
Orange Bowl, Advertising Very Rare..........................	1,500		2,000					
Bowl, 8¼", Ftd	48	60	67	65				850
Bowl, 8", Flat......................	42	50	58	48		850	600BO	650
Plate, 9", Ftd......................	80	115	160					1,200
Orange Basket, Very Rare		2,750						
Spittoon Whimsey, Rare	950							
GRAPE (IMPERIAL)								
Bowl, 10".............................	38	55	50				200IM	
Bowl, 5"...............................	18	26	24				35	
Fruit Bowl, 8¼"....................	36	50	48					265
Compote..............................	40	55	55				75AM	
Cup and Saucer	80		85				60	
Nappy..................................	26		40				30SM	
Tray, Center Handle	38							
Goblet, Rare	40	75	65				70AM	
Plate, 7"-12"	65	280	160	160			140	
Plate, 8½", Ruffled	58	80	75				80	
Pitcher................................	97	200	170				300SM	
Tumbler	15	38	30				155AQ	
Punch Bowl and Base..........	120	320	275				400AM	
Cup.....................................	18	25	20				30AM	
Water Bottle, Rare	125	190	170				270	
Milk Pitcher........................	265	325	300				185CM	
Basket, Handled, Rare..........	75		85				175SM	
Rose Bowl, Rare	175	195	180					
Decanter, w/stopper............	95	285	165				150	
Wine....................................	30	38	35				30	
Spittoon Whimsey................	975		2,050					
GRAPE (NORTHWOOD'S GRAPE AND CABLE)								
Bowl, 9"-10", Flat	65	75	80	75			125	
Bowl, 5½", Flat	30	37	40				50	
Scalloped Bowl, 5½"-11½"	40	48	45			750	58	
Bon Bon	35	45			295MO	500	760W	
Banana Boat, Ftd	170	220	210	250		3,200*	275	
Bowl, Ftd, 7"-9"	45	70	65	85				
Ice Cream Bowl, 11"	150	265	250				300W	
Orange Bowl, Ftd................	135	175	195			3,300	800	
Breakfast Set, 2 pcs............	125	160	155					
Candlesticks, pr	150	240	220				350	
Candlelamp, Complete.........	790	875	650					
Compote, Covered	2,650	410						
Compote, Open	350	475	490				750	
Sweetmeat, w/lid.................	875	250		1,400				
Sweetmeat Whimsey............	400	375						
Cookie Jar, w/lid	350	265				7,600	525W	
Centerpiece Bowl, Ftd.........	250	375	390				1,000AB	
Cup and Saucer, Rare.........	400	420						
Cologne, w/stopper	225	185	195					
Perfume, w/stopper.............	375	525						
Dresser Tray.......................	125	180	195				875W	
Pin Tray	105	130	145				675W	
Hatpin Holder.....................	195	165	180			3,100	350W	
Powder Jar w/lid	80	95	110	125				
Nappy..................................	45	58	65				80	
Fernery, Rare	1,250	900	980				2,000IB	
Hat......................................	35	45	40				80	
Ice Cream Sherbet..............	35	50	52				90	
Plate, 6"-9½", Flat...............	110	200	385	400		3,000	310	
Plate, Handgrip...................	55	70	76				120	
Plate, Ftd............................	60	80	85	115			1,000IG	
Plate, 2 Sides Up	55	70	75					
Shade.................................	200	175						
Punch Bowl and Base (Standard)	370	595	570	785			900	
Punch Bowl and Base (Small) ..	280	400	450	510			800	
Plate, Advertising.................			375					
Punch Bowl and Base (Banquet)	700	1,350	2,000	2,750		7,600	2,700	
Cup.....................................	25	35	40	45		290	60	
Butter	175	190	195				275	
Sugar, w/lid	80	110	115				170	
Creamer or Spooner	70	100	105				200IG	
Tobacco Jar, w/lid	160	370		1,200				
Pitcher, Standard	270	285	300				1,800IG	
Pitcher, Tankard..................	540	775	3,100				2,400IG	
Tumbler, Jumbo..................	54	60	75					

249

Grape Arbor (Northwood)

Grape and Cherry

Grape Delight

Grape Leaves (Millersburg)

	M	A	G	B	PO	AO	Pas	Re
Grape (Northwood's Grape and Cable) Cont'd								
Tumbler, Regular..................	40	55	70					
Decanter, w/stopper	650	800						
Shot Glass........................	225	150						
Spittoon, Rare	5,000	7,000	7,200					
Orange Bowl, Blackberry Interior, Rare		1,200						
Hatpin Holder Whimsey,Rare .		3,000	3,500					
Plate Whimsey, 14", Ftd.......		550						
GRAPE ARBOR (DUGAN)								
Bowl, 9½"-11", Ftd	50	75					110W	
GRAPE ARBOR (NORTHWOOD)								
Pitcher................................	225	540		2,600			3,800IG	
Tumbler	45	60		260			300IG	
Hat....................................	60			125			185W	
GRAPE AND CHERRY (ENGLISH)								
Bowl, 8½", Rare	65			150				
GRAPE DELIGHT (DUGAN)								
Rose Bowl, 6" Ftd	48	65		70			70W	
Nut Bowl, 6" Ftd	60	75		180			80W	
GRAPE FIEZE (NORTHWOOD)								
Bowl, 10½", Rare							450*IC	
GRAPE AND GOTHIC ARCHES (NORTHWOOD)								
Bowl, 10"..........................	48	55	75	60				
Bowl, 5"..........................	24	30	45	35				
Butter	115	185	125		390PL			
Sugar, w/lid	60	75	95	80			120PL	
Creamer or Spooner	60	75	95	80			120PL	
Pitcher.............................	200	375	585	360			700PL	
Tumbler	35	50	90	50			160PL	
GRAPE, HEAVY (DUGAN)								
Bowl, 5", Rare....................	145	155		375				
Bowl, 10"..........................	185	265		500				
GRAPE, HEAVY (IMPERIAL)								
Bowl, 9"..........................	40	50		56			60	
Bowl, 5"..........................	25	30		30			40	
Nappy..............................	35	38						
Plate, 8"..........................	55	80	85				190AM	
Plate, 6"..........................	50	60	65				80	
Plate, 11"..........................	240	300					275AM	
Fruit Bowl, w/base..............	295							
Custard Cup	20	35	35					
Punch Bowl, w/base............	200	450	440					
GRAPE LEAVES (MILLERSBURG)								
Bowl, Rare, 10"..................	575	800	900*				800V	
GRAPE LEAVES (NORTHWOOD)								
Bowl, 8¾"	60	70	75				275IB	
Brides Basket, Complete......		235						
GRAPE WREATH (MILLERSBURG)								
Bowl, 5"..........................	38	55	55				90CM	
Bowl, 7½"-9"	48	65	65	400				
Bowl, 10", Ice Cream	125	170	170					
Spittoon Whimsey, Rare	2,550	3,200						
GRAPEVINE LATTICE (DUGAN)								
Bowl, 8½"	40	57	55	60			70W	
Plate, 7"-9"	70	95		90			90W	
Bowl, 5"..........................	26						52W	
GRAPEVINE LATTICE (FENTON)								
Pitcher, Rare......................	320	580		625			850W	
Tumbler, Rare	52	75		90			100W	
GREEK KEY (NORTHWOOD)								
Bowl, 7"-8½"	70	90	100					
Plate, 9"-11", Rare	685	610	390	750		875		
Pitcher, Rare......................	765	1,000	1,085					
Tumbler, Rare	85	150	210					
GREEK KEY VT.								
Hatpin..............................		28						
GREENGARD FURNITURE (MILLERSBURG)								
Bowl, Rare..........................		800						
HAIR RECEIVER								
Complete	60							
HAMMERED BELL CHANDELIER								
Complete, 5 Shades.............							600W	
Shade, ea							95W	
HAND VASE								
One Shape, 5½"-8"..............	250	300						450*

Heavy Diamond

Heavy Hobnail

Heavy Prisms

Heavy Web

	M	A	G	B	PO	AO	Pas	Red
HANDLED TUMBLER								
One Size	50							
HANDLED VASE (IMPERIAL)								
One Shape	45							
HARVEST FLOWER (DUGAN)								
Pitcher, Rare	1,250							
Tumbler	120	350	400*					
HARVEST POPPY								
Compote	38	50			160		70	
HATCHET (U.S. GLASS)								
One Shape	145							
HATTIE (IMPERIAL)								
Bowl	38	58	65				60	
Rose Bowl	90							
Plate, Rare	275	650	530				900AM	
HAWAIIAN LEI (HIGBEE)								
Sugar	65							
Creamer	65							
HEADDRESS								
Bowl, 9", 2 Varieties	36		45	40				
Compote	40		50	46				
HEART BAND								
One Shape (Salt)	45							
HEART BAND SOUVENIR (McKEE)								
Mug, Small	85		95				110AO	
Mug, Large	90		110				120AO	
HEART AND HORSESHOE (FENTON)								
Bowl, 8½"	800							
Plate, 9", Rare	1,000							
HEART AND TREES (FENTON)								
Bowl, 8¾"	145		195	175				
HEART AND VINE (FENTON)								
Bowl, 8½"	38	46	42	40			58	
Plate, 9", Rare	195	375	215	200				
Spector Plate (Advertising), Rare	400							
Cooleemee Plate (Advertising), Rare	1,150							
HEARTS AND FLOWERS (NORTHWOOD)								
Bowl, 8½"	38	55	67	58			475IG	
Compote	136	150		256	625	1,200	1,200BO	
Plate, 9", Rare	170	900	1,625	275		1,650	420IB	
HEAVY DIAMOND								
Nappy	32							
HEAVY DIAMOND (IMPERIAL)								
Bowl, 10"	32							
Creamer	28							
Sugar	28							
Vase	42		48				55SM	
Compote	45		50					
HEAVY HEART (HIGBEE)								
Tumbler	75							
HEAVY HOBNAIL (FENTON)								
Vase, Rare		475*					375W	
HEAVY HOBS								
Lamp (Amber Base Glass)					195			
HEAVY PRISMS (ENGLISH)								
Celery Vase, 6"	76	90		85				
HEAVY SHELL (FENTON)								
Bowl, 8¼"							150	
Candleholder, ea							92	
HEAVY VINE								
Lamp	162							
Atomizer	65							
HEAVY WEB (DUGAN)								
Bowl, 10" Rare					800			
Plate, 11", Rare					1,095*			
HEINZ								
Bottle							36	
HEISEY								
Breakfast Set							185	
HEISEY CARTWHEEL								
Compote							65	
HEISEY FLORAL SPRAY								
Stemmed Candy, w/lid 11"							70IB	
HEISEY FLUTE								
Punch Cup	28							
HEISEY SET								
Creamer and Tray	75						165	

Heron

Hobnail

Hobnail Soda Gold

Hobstar and Cut Triangles

	M	A	G	B	PO	AO	Pas	Red
HEISEY #357								
Water Bottle	75							
Tumbler	42							
HERON (DUGAN)								
Mug, Rare	800*	265						
HEXAGON AND CANE (IMPERIAL)								
Covered Sugar	65							
HEX-OPTIC (JEANETTE)								
Tumbler							40CL	
HICKMAN								
Castor Set, 4 pc	295						450	
HOBNAIL (FENTON)								
Vase, 5"-11"							75W	
HOBNAIL (MILLERSBURG)								
Pitcher, Rare	1,800	1,900	2,150	1,600				
Tumbler, Rare	775	500	1,000	950				
Rose Bowl, Scarce	175	350	450					
Spittoon, Rare	600	600	1,600					
Butter, Rare	470	600	650	900				
Sugar, w/lid, Rare	350	500	570	780				
Creamer or Spooner, Rare	275	375	450	500				
HOBNAIL VT. (MILLERSBURG)								
Vase Whimsey, Rare	250*	250*	250*					
Rose Bowl, Rare	850*							
Jardinere, Rare		900*	1,000*					
HOBNAIL, MINIATURE								
Tumbler, 2½"	48							
Pitcher 6", Rare	210*							
HOBNAIL PANELS (McKEE)								
Vase, 8¼"							65CM	
HOBNAIL SODA GOLD (IMPERIAL)								
Spittoon, Large	48		65				50W	
HOBSTAR (IMPERIAL)								
Bowl, 10", Berry	35						45	
Bowl, 5", Berry	24						35	
Bowls, 6"-12", Various Shapes	28						40	
Fruit Bowl, w/base	50	85	75					
Cookie Jar, w/lid	58		95					
Butter	65	195	185				80CM	
Sugar, w/lid	45	90	85				50CM	
Creamer or Spooner	40	80	75				45CM	
Pickle Castor, Complete	375							
Bride's Basket, Complete	75							
HOBSTAR AND ARCHES (IMPERIAL)								
Bowl, 9"	48	60	57				52SM	
Fruit Bowl w/base	58	75	70					
HOBSTAR BAND (IMPERIAL)								
Celery	85							
Compote, Rare	90							
Bowl, Rare	70							
Pitcher, 2 Shapes, Rare	275							
Tumbler, 2 Shapes, Rare	54							
HOBSTAR AND CUT TRIANGLES (ENGLISH)								
Rose Bowl	45	55	70					
Bowl	30	42	58					
Plate	65	95	110					
HOBSTAR AND FEATHER (MILLERSBURG)								
Punch Bowl and Base (Open), Rare	1,800		2,800					
Punch Bowl and Base (Tulip), Rare		6,000						
Punch Cup, Scarce	26	35		80				
Rose Bowl, Giant, Rare	3,000*	1,700	2,000					
Vase Whimsey, Rare		4,000	4,000					
Punch Bowl Whimsey, Rare			7,500					
Compote Whimsey (From Rose Bowl), Rare		7,500						
Bowl, 5" (Round), Rare		450						
Bowl, 5" (Diamond), Rare	400							
Bowl, 5" (Heart), Rare	300							
Butter, Rare	1,200	1,500	1,500					
Sugar w/lid, Rare	800	900	900				3,000V	
Creamer, Rare	700	800	800					
Spooner, Rare	700	800	800					
Dessert, Stemmed, Rare	650							
Compote, 6" Rare	1,500							

Hobstar Flower

Holly Sprig or Whirl

Honeycomb and Clover

Honeycomb and Hobstar

	M	A	G	B	PO	AO	Pas	Red
HOBSTAR AND FILE								
Pitcher, Rare	1,500							
Tumbler, Rare	170*							
HOBSTAR FLOWER (NORTHWOOD)								
Compote, Scarce	54	65	70	75				
HOBSTAR AND FRUIT (WESTMORELAND)								
Bowl, 6", Rare					90	270		
Bowl, 10" Rare					125			
Plate, 10½", Rare							350IB	
HOBSTAR PANELS (ENGLISH)								
Creamer	45							
Sugar, Stemmed	45							
HOBSTAR REVERSED (ENGLISH)								
Spooner	45							
Butter	58	75		70				
Frog and Holder	50							
HOBSTAR WHIRL (WHIRLAGIG)								
Compote, 4½"	48	58		60				
HOLIDAY								
Bottle							67	
HOLIDAY (NORTHWOOD)								
Tray, 11" Rare	350							
HOLLY (FENTON)								
Bowl, 8"-10"	48	60	70	70		390	265V	750
Compote, 5"	32	38	45	36			55	495
Goblet	30	36					58	550
Hat	28	32	38	30			45	550
Plate, 9"	110	195	400	300		820	175	2,000
HOLLY AND BERRY (DUGAN)								
Bowl, 7"-9"	38	45	50	48	60			
Nappy	42	55	58	50	65			
Gravy Boat, Handled		58		62	135			
HOLLY, PANELLED (NORTHWOOD)								
Bowl		70	65					
Bon Bon, Ftd	55	90	80					
Creamer or Sugar	475*							
Pitcher, Rare		11,000*						
HOLLY SPRIG VT. (MILLERSBURG)								
Bowl, Scarce	190		245					
HOLLY SPRIG OR WHIRL (MILLERSBURG)								
Deep Sauce, Rare	175	175	200					
Nappy, Tri-Cornered, Rare	125	195	180					
Bon Bon (Plain)	55	60	58					
Bon Bon (Isaac Benesch), Rare	165							
Bowl, 7"-10", Round or Ruffled	50	58	60				55CM	
Compote, Very Rare	450	625					1,000V	
Rose Bowl Whimsey, Rare							1,250V	
Bowl, Tri-Cornered, 7"-10"	150	195	185					
HOLM SPRAY								
Atomizer, 3"	45							
HOMESTEAD								
Shade	40							
HONEYCOMB (DUGAN)								
Rose Bowl	190				235			
HONEYCOMB AND CLOVER (FENTON)								
Bon Bon	38	50	57	48				
Compote	28	50	57	48			70AM	
HONEYCOMB AND CLOVER (IMPERIAL-FENTON)								
Spooner, Rare	90							
HONEYCOMB AND HOBSTAR (MILLERSBURG)								
Vase, 8¼", Rare		6,000		6,100				
HONEYCOMB ORNAMENT								
Hatpin		70		70				
HORN OF PLENTY								
Bottle	56							
HORN, POWDER (CAMBRIDGE)								
Candy Holder	185							

Hot Springs

Imperial Basket

Inca

Intaglio Daisy

	M	A	G	B	PO	AO	Pas	Red
HORSES HEADS (FENTON)								
Bowl, 7½", Flat	67		90	90			210W	1,100
Bowl, 7"-8", Ftd	80		95	95			220V	1,100
Plate, 6½"-8½"	100		210	510				
Rose Bowl, Ftd	120		140	140			295	
HORSESHOE								
Shot Glass	42							
HOT SPRINGS SOUVENIR								
Vase, 9⅞", Rare	115							
HOURGLASS								
Bud Vase	46							
HUMPTY-DUMPTY								
Mustard Jar	75							
HYACINTH								
Lamp	1,900							
ICE CRYSTALS								
Bowl, Ftd							85	
Candlesticks, pr							160	
Salt, Ftd							65	
IDYLL (FENTON)								
Vase, Rare		675		650				
ILLINOIS DAISY (ENGLISH)								
Bowl, 8"	40							
Cookie Jar, w/lid	55							
ILLUSION (FENTON)								
Bon Bon	47			65				
Bowl	55			90				
IMPERIAL BASKET (IMPERIAL)								
One Shape, Rare	48						70SM	
IMPERIAL DAISY (IMPERIAL)								
Shade	38							
IMPERIAL GRAPE (IMPERIAL)								
Shade	65							
IMPERIAL #5 (IMPERIAL)								
Bowl, 8"	40							
Vase, 6", Rare	90						45AM	
IMPERIAL #9 (IMPERIAL)								
Compote	38							
IMPERIAL PAPERWEIGHT (IMPERIAL)								
Advertising Weight, Rare		900*						
INCA								
Vase, 7", Rare	750*	900*						
INDIANA GOBLET (INDIANA GLASS)								
One Shape, Rare							800AM	
INDIANA STATEHOUSE (FENTON)								
Plate, Rare	2,850			3,350				
INSULATOR (VARIOUS MAKERS)								
Various Sizes	25+							
INTAGLIO DAISY (ENGLISH)								
Bowl, 7½"	48							
Bowl, 4½"	26							
INTAGLIO FEATHERS								
Cup	25							
INTAGLIO OVALS (U.S. GLASS)								
Bowl, 7"						65		
Plate, 7½"						80		
INTERIOR PANELS								
Mug	75							
INTERIOR POINSETTIA (NORTHWOOD)								
Tumbler, Rare	465							
INTERIOR RAYS								
Sherbet	35							
INTERIOR RAYS (WESTMORELAND)								
Covered Butter	65							
Sugar, Creamer or Jam Jar, ea	40							
INTERIOR SWIRL								
Vase, 9" Ftd	37							
Spittoon					95			

Inverted Feather

Inverted Strawberry

Inverted Thistle

Iris

	M	A	G	B	PO	AO	Pas	Red
INVERTED COIN DOT (NORTHWOOD-FENTON)								
Pitcher	325	450		400				
Tumbler	75	95		85				
Bowl	38	45	67	50			70	
Rose Bowl	47		60				85	
INVERTED FEATHER (CAMBRIDGE)								
Cracker Jar, w/lid		1,000	585					
Covered Butter, Rare	450	495						
Sugar, Creamer or Spooner, Rare	390	320						
Pitcher, Tall, Rare	4,000							
Tumbler, Rare	500		600					
Compote	100							
Punch Bowl, w/base, Rare	2,800		3,800*					
Cup, Rare	60		85					
Wine, Rare	190							
Squat Pitcher, Rare	1,000							
INVERTED STRAWBERRY								
Bowl, 9"-10½"	90	290	285	325				
Bowl, 5"	38	50	47					
Sugar, Creamer or Spooner, ea, Rare	90			150				
Covered Butterdish		750						
Candlesticks, pr, Rare	300	425	400					
Compote, Large, Rare	300	500	450					
Compote, Small, Rare	400			350				
Powder Jar, Rare	175		230					
Ladies Spittoon, Rare	600	700	650					
Milk Pitcher, Rare		1,400		*				
Pitcher, Rare	2,000	2,900	2,800					
Tumbler, Rare	325	265	375					
Table Set, 2 pcs, Rare (Stemmed)		1,275		1,000				
Celery, Rare		1,275	1,350	1,350				
Compote Whimsey, Rare	475		550					
INVERTED, THISTLE (CAMBRIDGE)								
Bowl, 9", Rare		300	300					
Spittoon, Rare		3,900*						
Covered Box, Rare					350*			
Pitcher, Rare	3,700	3,500						
Tumbler, Rare	425	350						
Butter, Rare	500	600	700					
Sugar, Creamer or Spooner, Rare	350	380	480					
Chop Plate, Rare		2,450						
Milk Pitcher, Rare		2,600						
Bowl, 5", Rare		185	195					
IOWA								
Small Mug, Rare	85							
IRIS (FENTON)								
Compote	47	60	58	75			275W	
Buttermilk Goblet, Scarce	52	67	65				78AM	
IRIS, HEAVY (DUGAN)								
Pitcher	475	850			1,250		1,200W	
Tumbler	85	85					160W	
IRIS HERRINGBONE (JEANETTE)								
Various Shapes, Prices range from $3.00 to $16.00 each in this late pattern.								
ISAAC BENESCH								
Advertising Bowl, 6½"		350						
I.W. HARPER								
Decanter, w/stopper	75							
JACK-IN-THE-PULPIT (DUGAN)								
Vase	45	75		80	95			
JACKMAN								
Whiskey Bottle	30							
JACOB'S LADDER								
Perfume	47							
JACOB'S LADDER VT. (U.S. GLASS)								
Rose Bowl	52							

	M	A	G	B	PO	AO	Pas	Re
JACOBEAN RANGER (CZECHOSLOVAKIAN AND ENGLISH)								
Pitcher	250							
Tumbler	75							
Juice Tumbler	70							
Miniature Tumbler	65							
Decanter w/stopper	150							
Wine	28							
Bowls, Various Sizes	30+							
JARDINERE (FENTON)								
Various Shapes	350	500	500	450				
JELLY JAR								
Complete, Rare	65							
JEWEL BOX								
Ink Well	75							
JEWELED HEART (DUGAN)								
Bowl, 10"		95			135			
Bowl, 5"		42			65			
Pitcher, Rare	775							
Tumbler, Rare	100						575W	
Plate, 6"	125				210			
JEWELS (IMPERIAL)								
Candlesticks, pr	90	140					70	310
Bowls, Various Sizes		45					80	345
Hat Shape		65					86	290
Vase	90	150					80	270
Creamer or Sugar		70					70	235
JOCKEY CLUB (NORTHWOOD)								
Bowl, 7"		245						
KANGAROO (AUSTRALIAN)								
Bowl, 9½"	170	197						
Bowl, 5"	47	54						
KEYHOLE (DUGAN)								
Exterior Pattern of Raindrops Bowls								
KINGFISHER AND VARIANT (AUSTRALIAN)								
Bowl, 5"	42	50						
Bowl, 9½"	170	190						
KITTEN								
Miniature Paperweight, Rare	250							
KITTENS								
Bottle							52	
KITTENS (FENTON)								
Bowl, 4", Scarce	118	250		195			245V	
Bowl, 2 Sides Up, Scarce	116			210				
Cup and Saucer, Scarce	252			640				
Spooner, 2½", Rare	147			265			210V	
Plate, 4½", Scarce	128			170				
Cereal Bowl, Scarce	129			165				
Vase, 3"	160			215			230V	
Spittoon Whimsey, Rare	3,050			3,500				
KIWI (AUSTRALIAN)								
Bowl, 10", Rare	350	300						
KNIGHT TEMPLAR (NORTHWOOD)								
Advertising Mug, Rare	800						1,000	
KNOTTED BEADS (FENTON)								
Vase, 4"-12"	28		40	37			85V	
KOKOMO (ENGLISH)								
Rose Bowl, Ftd	45		60	50				
KOOKABURRA AND VTS. (AUSTRALIAN)								
Bowl, 5"	42	48						
Bowl, 10"	180	192						
LACY DEWDROP (WESTMORELAND)								
Pitcher							585	
Compote, Covered							285	
Bowl, Covered							240	
Banana Boat							350	
Tumbler							250	
Goblet							150	
Creamer							90	
Sugar							90	
(Note: All Items Listed Are in Pearl Carnival)								
LADY'S SLIPPER								
One Shape, Rare	250							

Jeweled Heart

Jewels

Kingfisher Vt.

Lacy Dewdrop

Lattice and Daisy

Leaf and Beads

Leaf Swirl

Lily of the Valley

	M	A	G	B	PO	AO	Pas	Red
LARGE KANGAROO (AUSTRALIAN)								
Bowl, 5"..........................	40	52						
Bowl, 10"........................	165	185						
LATTICE (DUGAN)								
Bowl, Various Sizes	56	65					80	
LATTICE AND DAISY (DUGAN)								
Bowl, 9"...........................	60							
Bowl, 5"...........................	30							
Pitcher.............................	235			285			320W	
Tumbler	40	55		48			58W	
LATTICE AND GRAPE (FENTON)								
Pitcher.............................	260	425	485	410	1,800		650W	
Tumbler	38	45	55	40	375		200W	
Spittoon Whimsey, Rare	2,350							
LATTICE HEART (ENGLISH)								
Bowl, 10".........................	50	75		70				
Bowl, 5"..........................	30			40				
Compote...........................	60	90		85				
LATTICE AND LEAVES								
Vase, 9½"........................	50			67				
LATTICE AND POINTS (DUGAN)								
Vase	35	42					55W	
LATTICE AND PRISMS								
Cologne, w/stopper	55							
LATTICE AND SPRAYS								
Vase, 10½".........................	40							
LAUREL								
Shade...............................							40	
LAUREL BAND								
Tumbler	42							
LAUREL AND GRAPE								
Vase, 6"...........................	110							
LAUREL LEAVES (IMPERIAL)								
Plate...................................	47	55					60SM	
LBJ HAT								
Ashtray	26							
LEA AND VT. (ENGLISH)								
Bowl, Ftd...........................	38							
Pickle Dish, Handled	40							
Creamer, Ftd	40	52						
LEAF AND BEADS (NORTHWOOD-DUGAN)								
Bowl, 9"..............................	70	80	85					
Plate Whimsey....................	200							
Candy Dish, Ftd	45	54	52		290		210	
Rose Bowl, Ftd	50	65	67	75	345	400	500W	
Nut Bowl, Rare						550		
(Note: Add 25% for Patterned Interior)								
LEAF CHAIN (FENTON)								
Bon Bon	35	48	54	46				
Bowl, 7"-9"	42	52	58	48		1,200	65W	650
Plate, 7½"				90				
Plate, 9¼"	65	92	100	85		2,100		
LEAF COLUMN (NORTHWOOD)								
Vase	35	45	40		156		170IB	
Shade.................................							85	
LEAF AND LITTLE FLOWERS (MILLERSBURG)								
Compote, Miniature, Rare....	350	475	425					
LEAF RAYS (DUGAN)								
Nappy, Either Exterior.........	28	37			42		40W	
LEAF SWIRL (WESTMORELAND)								
Compote............................	50	55		58TL			60AM	
LEAF SWIRL AND FLOWER (FENTON)								
Vase	45						65W	
LEAF TIERS (FENTON)								
Bowl, 10", Ftd....................	58							
Bowl, 5", Ftd.....................	30							
Butter, Ftd.........................	175							
Sugar, Ftd	88							
Creamer, Spooner, Ftd........	82							
Pitcher, Ftd, Rare	460	650	630	685				
Tumbler, Ftd, Rare	76	90	87	95				
Banana Bowl Whimsey	190							
LILY OF THE VALLEY (FENTON)								
Pitcher, Rare.......................				4,500				
Tumbler, Rare	600			450				

Little Beads

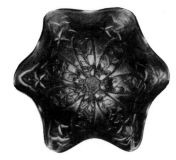

Little Stars

Loganberry

Long Thumbprint

	M	A	G	B	PO	AO	Pas	Re
LINED LATTICE (DUGAN)								
Vase, 7"-14"	35	48	45	48	168		60W	
LION (FENTON)								
Bowl, 7", Scarce	85			150				
Plate, 7½", Rare	500							
LITTLE BARREL (IMPERIAL)								
One Shape	175		195				210AM	
LITTLE BEADS								
Bowl, 8"	18				45			
Compote, Small	26				48	85	40AQ	
LITTLE DAISIES (FENTON)								
Bowl, 8"-9½", Rare	290			345				
LITTLE DAISY								
Lamp, 8",								
Complete							395	
LITTLE DARLING								
Bottle	52							
LITTLE FISHES (FENTON)								
Bowl, 10", Flat or Ftd	180		190	265			850W	
Bowl, 5½", Flat or Ftd	40	75	56	70			90V	
Plate, 10½", Rare				600			950W	
LITTLE FLOWERS (FENTON)								
Bowl, 9¼"	48	90	68	85			90AM	1,67?
Bowl, 5½", Rare	28	38	32	36			50V	
Plate, 7", Rare	185							
Plate, 10", Rare	750							
LITTLE MERMAID								
One Shape							75	
LITTLE OWL								
Hatpin, Rare	115		150	165			175W	
LITTLE STARS (MILLERSBURG)								
Bowl, 4", Rare			375					
Bowl, 7", Scarce	65	70	90	1,350				
Bowl, 9", Rare	450	550	575					
Bowl, 10½", Rare	585	700	750					
Plate, 7⅝", Rare			475					
LITTLE SWAN								
Miniature, 2"							75	
LOG								
Paperweight, 3"X1¼", Rare	150							
LOGANBERRY (IMPERIAL)								
Vase, Scarce	225	485	375				450AM	
LONG HOBSTAR								
Bowl, 8½"	45							
Bowl, 10½"	56							
Compote	65							
Punch Bowl and Base	125						135CM	
LONG HORN								
Wine	35							
LONG LEAF (DUGAN)								
Bowl, Ftd					158			
LONG PRISMS								
Hatpin		35						
LONG THUMBPRINT (DUGAN)								
Vase, 7"-11"	28	33	36	38	146			
Bowl, 8¾"	30	38						
Compote	35	39	39					
Creamer, Sugar, ea	38						50SM	
Butter	60							
LOTUS AND GRAPE (FENTON)								
Bon Bon	36	40	45	42				85(
Bowl, 7", Flat	45	48	50	52				
Bowl, 7", Ftd	48	50		56				
Plate, 9½", Rare	175	450	625	500				
LOTUS LAND (NORTHWOOD)								
Bon Bon	595							
LOUISA (WESTMORELAND)								
Rose Bowl	52	65	70	65			140AM	
Candy Dish, Ftd	48	60	65				70AQ	
Bowl, Ftd		50	50RG				50AQ	
Plate, 8", Ftd, Rare		150						
LOVEBIRDS								
Bottle, w/stopper	400							
LUCKY BANK								
One Shape	38							
LUCKY BELL								
Bowl, 8¾", Rare	80							

258

Lustre Flute

Lustre Rose

Many Fruits

Maple Leaf

	M	A	G	B	PO	AO	Pas	Red
LUSTER								
Tumbler	40							
LUSTRE AND CLEAR (FENTON)								
Fan Vase	40		55	50			60IG	
LUSTRE AND CLEAR (IMPERIAL)								
Pitcher	195							
Creamer or Sugar	36	65					60	
Compote	40							
Celery Tray, 8"	35						30SM	
Bowl, 5"	20						30	
Bowl, 10"	36						50	
Vase, 8", Ftd	85	110	120					
Shakers, pr	70							
Tumbler	40							
Rose Bowl	50							
Butter Dish	65							
Wall Vase	30							
LUSTRE AND CLEAR (LIGHTOLIER)								
Shade	40							
LUSTRE FLUTE (NORTHWOOD)								
Bowl, 5½"	36							
Bowl, 8"	38	55	54					
Bon Bon		58	55					
Creamer or Sugar	35	55	52					
Hat	28	40	36					
Compote		48	45					
Sherbet	26							
Nappy	30	42	40					
Punch Bowl and Base	145	160	150					
Cup	15	24	20					
LUSTRE ROSE (IMPERIAL)								
Bowl, 7"-11", Flat	35	45	48				60CM	
Bowl, 9"-12", Ftd	38	56	58				75	2,250
Fernery	40	60	65	75			95AM	
Plate, 6"-9"	56	70	75				85	
Butter	58	74	72				100AM	
Sugar	45	56	54				80AM	
Creamer or Spooner	42	55	55				70AM	
Berry Bowl, 8"-9"	38	42	46				40	
Berry Bowl, 5"	20	26	28				20	
Pitcher	85	97	97				160AM	
Rose Bowl	60	70	80				70CL	
Tumbler	25	30	28				50AM	
Milk Pitcher	57		67				120AM	
Plate Whimsey, Ftd	48		58				65CM	
LUTZ (McKEE)								
Mug, Ftd	45							
MAGNOLIA DRAPE								
Pitcher	240							
Tumbler	50							
MAGPIE (AUSTRALIAN)								
Bowl, 6"-10"	45	56						
MAIZE (LIBBEY)								
Vase, Celery, Rare							185CL	
Syrup, Rare							235CL	
MALAGA (DUGAN)								
Bowl, 9" Scarce	70	90						
Plate, 10", Rare		250					270AM	
MANY FRUITS (DUGAN)								
Punch Bowl, w/base	250	450					1,200W	
Cup	25	30	35	40			40W	
MANY PRISMS								
Perfume, w/stopper	65							
MANY STARS (MILLERSBURG)								
Bowl, 9", Ruffled, Scarce	300	530	460	2,000			1,750V	
Bowl, 9½", Round, Rare	325	575	495					
Tri-cornered Bowl, Rare		1,500						
MAPLE LEAF (DUGAN)								
Bowl, 9" Stemmed	70	110		90				
Bowl, 4½", Stemmed	28	35	40	30				
Butter	100	125		118				
Sugar	65	75		70				
Creamer or Spooner	55	65		62				
Pitcher	150	325		300				
Tumbler	30	48		40				

	M	A	G	B	PO	AO	Pas	Re

MARIE
(FENTON)
    Rustic Vase Interior
    Base Pattern

MARILYN
(MILLERSBURG)

	M	A	G	B	PO	AO	Pas	Re
Pitcher, Rare	700	975	1,300					
Tumbler, Rare	145	280	400					

MARY ANN
(DUGAN)

	M	A	G	B	PO	AO	Pas	Re
Vase, 7", 2 Varieties	35	80						
Loving Cup, 3 Handles, Rare	250*							

MASSACHUSETTS
(U.S. GLASS)

	M	A	G	B	PO	AO	Pas	Re
Vase	150*							

MAY BASKET
(ENGLISH)

	M	A	G	B	PO	AO	Pas	Re
Basket, 7½"			80					
Bowl, 9", Rare			140					

MAYFLOWER

	M	A	G	B	PO	AO	Pas	Re
Bowl, 7½"	28	40			156		50	
Compote	38	50					60	
Shade	30							
Hat	32	36			150		60	

MAYFLOWER
(MILLERSBURG)
    Exterior Pattern on Grape
    Leaves Bowls

MAYPOLE

	M	A	G	B	PO	AO	Pas	Re
Vase, 6¼"	42	50	56					

MELON RIB
(IMPERIAL)

	M	A	G	B	PO	AO	Pas	Re
Candy Jar, w/lid	30							
Pitcher	60							
Tumbler	24							
Powder Jar, w/lid	35							
Shakers, pr	50							

MEMPHIS
(NORTHWOOD)

	M	A	G	B	PO	AO	Pas	Re
Bowl, 10"	165	195	200					
Bowl, 5"	30	45	50					
Fruit Bowl, w/base	430	540	585	2,250			3,000IB	
Punch Bowl, w/base	395	490	550				3,000IG	
Cup	30	38	42				60IB	
Sugar or Creamer		60						

MIKADO (FENTON)

	M	A	G	B	PO	AO	Pas	Re
Compote, Large	195		645	385			550W	5,000

MILADY
(FENTON)

	M	A	G	B	PO	AO	Pas	Re
Pitcher	475	650	650	500				
Tumbler	95	140	120	100				

MINIATURE BELL

	M	A	G	B	PO	AO	Pas	Re
Paperweight, 2½"	45							

MINIATURE FLOWER BASKET
(WESTMORELAND)

	M	A	G	B	PO	AO	Pas	Re
One Shape	60							

MINIATURE HOBNAIL

	M	A	G	B	PO	AO	Pas	Re
Cordial Set, Rare	750							

MINIATURE INTAGLIO
(WESTMORELAND)

	M	A	G	B	PO	AO	Pas	Re
Nut Cup, Stemmed, Rare	450						550W	

(Note: Also Known as "Wild
Rose Wreath")

MINIATURE SHELL

	M	A	G	B	PO	AO	Pas	Re
Candleholder, ea							75CL	

MIRRORED PEACOCKS

	M	A	G	B	PO	AO	Pas	Re
Tumbler, Rare	150*							

MIRRORED LOTUS
(FENTON)

	M	A	G	B	PO	AO	Pas	Re
Bon Bon	50		60	57				
Bowl, 7"-8½"	45		58	50				
Plate, 7½", Rare	145			185			2,200*IB	
Rose Bowl, Rare	90			145			850W	

MITERED DIAMOND AND
PLEATS (ENGLISH)

	M	A	G	B	PO	AO	Pas	Re
Bowl, 4½"	20			30				
Bowl, 8½", Shallow	35			40				

MITERED OVALS
(MILLERSBURG)

	M	A	G	B	PO	AO	Pas	Re
Vase, Rare	5,000	4,700	4,500					

Marilyn

Massachusetts

Memphis

Mitered Ovals

	M	A	G	B	PO	AO	Pas	Red
MOON AND STAR (WESTMORELAND)								
Compote (Pearl Carnival)							365	
MOONPRINT (ENGLISH)								
Bowl, 8¼"	45							
Candlesticks, Rare, ea	50							
Compote	45							
Jar, w/lid	60			85				
Vase	50							
Cheese Keeper, Rare	135							
Sugar, Stemmed	50							
Banana Boat, Rare	135							
Creamer	45							
Butter	100							
Bowl, 14"	80							
Milk Pitcher, Scarce	100							
MORNING GLORY (IMPERIAL)								
Vase, 8"-16"	40+	90+	70+				80+SM	
MORNING GLORY (MILLERSBURG)								
Pitcher, Rare	7,000	8,500	9,700					
Tumbler, Rare	1,000	975	1,000					
MOXIE								
Bottle, Rare							78	
MULTI-FRUITS AND FLOWERS (MILLERSBURG)								
Dessert, Stemmed, Rare		750	750					
Punch Bowl, w/base, Rare	2,000	2,300	2,600	4,000				
Cup, Rare	40	45	50	80				
Pitcher, Rare (Either Base)	8,800	7,500	8,000					
Tumbler, Rare	1,000	950	850					
MY LADY								
Powder Jar w/lid	87							
MYSTIC (CAMBRIDGE)								
Vase, Ftd, Rare	120							
NAPOLEON								
Bottle							70	
NARCISSUS AND RIBBON (FENTON)								
Wine Bottle, w/stopper, Rare	885							
NAUTILUS (DUGAN-NORTHWOOD)								
Lettered, Rare		285			320			
Unlettered		185			250			
Giant Compote, Rare	2,600							
Vase Whimsey, Rare	1,600*	1,600*						
NEAR CUT (CAMBRIDGE)								
Decanter w/stopper, Rare			2,600					
NEAR CUT SOUVENIR (CAMBRIDGE)								
Mug, Rare	175							
Tumbler, Rare	210							
NEAR CUT WREATH (MILLERSBURG)								
Exterior Pattern Only								
NELL (HIGBEE)								
Mug	65							
NESTING SWAN (MILLERSBURG)								
Rose Bowl, Rare	2,500							
Bowl, 10" Scarce, Round or Ruffled	265	370	425	2,800			2,650V	
Spittoon Whimsey, Rare			4,700*					
Bowl, Tri-Cornered, Rare	750	1,200	1,200				750CL	
NEW ORLEANS SHRINE (U.S. GLASS)								
Champagne							90CL	
NIGHT STARS (MILLERSBURG)								
Bon Bon, Rare	650	650	580RG					
Card Tray, Rare		700					1,100V	
Nappy, Tri-Cornered, Very Rare		1,800						
NIPPON (NORTHWOOD)								
Bowl, 8½"	48	56	56	60		745	625IG	
Plate, 9"	300	375	375	350			600W	

Morning Glory

Nautilus

Near Cut

Nippon

Nu-Art Chrysanthemum

Number 270

Octagon

Octet

	M	A	G	B	PO	AO	Pas	Red
NORRIS N. SMITH (NORTHWOOD)								
Advertising Bowl, 5¼"		195						
NORTHERN STAR (FENTON)								
Card Tray, 6"	38							
Bowl, 6"-7"	28							
Plate, 6½", Rare	70							
NORTHWOOD JACK-IN THE PULPIT								
Vase, Various Sizes	36	48	45	50	90	210	75	
NORTHWOOD JESTER'S CAP								
Vase	40	50	54	58	75	215	70	
NORTHWOOD'S LOVELY								
Bowl, 9" (Leaf and Beads Exterior), Rare	600	600						
NORTHWOOD'S NEARCUT								
Compote	85	130						
Goblet, Rare	110	145						
Pitcher, Rare	1,500							
NORTHWOOD'S POPPY								
Bowl, 7"-8¾"	48	56		58	170	370		
Pickle Dish, Oval	48	58		65		800	80	
Tray, Oval, Rare		180	225				200	
NU-ART (IMPERIAL)								
Plate, Scarce	500	695	850	1,095			750W	
Shade	42							
NU-ART CHRYSANTHEMUM (IMPERIAL)								
Plate, Rare	700	995	1,000			1,200AM		
NUGGATE								
Pitcher 6"				90				
NUMBER 4 (IMPERIAL)								
Bowl, Ftd	26						36SM	
Compote	30							
NUMBER 270 (WESTMORELAND)								
Compote		70	125RG		85		115AQ	
NUMBER 2176 (SOWERBY)								
Lemon Squeezer	50							
NUMBER 2351 (CAMBRIDGE)								
Bowl, 9", Rare			285*					
Punch Cup		65	65					
OCTAGON (IMPERIAL)								
Bowl, 8½"	38							
Bowl, 4½"		30	28					
Goblet	65						80	
Butter	85		126					
Sugar	58		76					
Creamer or Spooner	48		65					
Toothpick Holder, Rare	65	85						
Pitcher	120	400	225				195	
Tumbler	28	50	44				45	
Decanter, Complete	85	250						
Wine	25	50						
Vase, Rare	85	130	125					
Milk Pitcher, Scarce	150	210	170				190	
Cordial							65	
Shakers, pr, Old Only		150						
OCTET (NORTHWOOD)								
Bowl, 8½"	48	60					80W	
OHIO STAR (MILLERSBURG)								
Vase, Rare	1,500	1,850	1,600			12,000*	3,500W	
Compote, Rare	950							
Vase Whimsey, Rare		1,500	1,500			10,000*		
OKLAHOMA (MEXICAN)								
Tumble-Up, Complete	175							
Pitcher, Rare	495							
Tumbler, Rare	450							
Shade, Various Sizes, Rare							87CL	
OLYMPIC (MILLERSBURG)								
Compote, Small, Rare		3,000	3,000					
OLYMPUS								
Shade	50							

Optic Flute

Orange Peel

Orange Tree

Orange Tree and Scroll

	M	A	G	B	PO	AO	Pas	Red
OMNIBUS								
Tumbler, Rare	375*							
OPEN FLOWER (DUGAN)								
Bowl, 7", Flat or Ftd............	28	36	38		85			
OPEN ROSE								
(IMPERIAL)								
Fruit Bowl, 7"-10"...............	38	65	68				60	
Bowl, 9"-12", Ftd	45	42	46				90	
Bowl, 9", Flat.....................	38	40	42				60	
Bowl, 5½", Flat	20	28	28				40	
Plate, 9"............................	100	290	185				200AM	
OPTIC (IMPERIAL)								
Bowl 9".............................		75						
Bowl 6".............................		47						
OPTIC AND BUTTONS								
(IMPERIAL)								
Bowls, 5"-8"......................	28							
Bowl, Handled, 12"	42							
Pitcher, Small, Rare	185							
Plate, 10½"	70							
Tumbler, 2 Shapes,								
Rare	95							
Goblet	58							
Cup and saucer, Rare	185							
OPTIC FLUTE								
(IMPERIAL)								
Bowl, 10"...........................	40	80					50SM	
Bowl, 5".............................	25	40					28SM	
Compote............................	55							
OPTIC 66								
(FOSTORIA)								
Goblet	45							
ORANGE PEEL								
(WESTMORELAND)								
Punch Bowl, w/base............	225						265TL	
Cup	20	32					35TL	
Custard Cup, Scarce	26							
Dessert, Stemmed, Scarce ...	37	65	70RG				75TL	
ORANGE TREE								
(FENTON)								
Butter	236			250			290W	
Bowl, 8"-10", Flat	38			60			375CB	695
Bowl, 9"-11", Ftd	75		210	120			130	
Ice Cream, w/stem, Small ...	24			28				
Bowl, 5½", Ftd	28		30	32			875IB	
Breakfast Set, 2 Pieces	190		230	220			270W	
Plate, 8"-9½"......................	98			210	2,000		210W	
Powder Jar, w/lid	70		400	87			120	
Mug, 2 Sizes......................	56	87	89	92			170AM	575
Loving Cup	250		290	200	4,500*	5,000	500W	
Punch Bowl, w/base............	200		325	300	1,000MO		400	
Cup	28		36	32			38	
Compote, Small	32	38	50	36			45	
Goblet, Large	90							
Wine.................................	28					60	70AQ	
Sugar	65			75			150W	
Pitcher, 2 Designs...............	210			300			8,000LO	
Tumbler	37			42			75W	
Creamer or Spooner	45			65			110W	
Rose Bowl..........................	48	58	60	54			250W	800
Hatpin Holder.....................	126		190	180	2,350		275W	
Hatpin Holder Whimsey,								
Rare				2,500				
Centerpiece Bowl, Rare........	875		1,000	1,600				
Cruet Whimsey, Rare...........				1,200				
Orange Bowl Whimsey, 12"..			500					
ORANGE TREE AND SCROLL								
(FENTON)								
Pitcher..............................	465		590	585				
Tumbler	50		85	80				
ORANGE TREE ORCHARD								
(FENTON)								
Pitcher..............................	420	520	525	500			675W	
Tumbler	45	48	52	46			125W	
ORIENTAL POPPY								
(NORTHWOOD)								
Pitcher..............................	500	700	950	2,500			1,800IG	
Tumbler	38	45	58	265			195IB	
OSTRICH								
(AUSTRALIAN)								
Compote, Large, Rare	125	160						
Cake Stand, Rare	260	320						

Oval and Round

Palm Beach

Panelled Dandelion

Panelled Smocking

	M	A	G	B	PO	AO	Pas	Red
OVAL AND ROUND (IMPERIAL)								
Plate, 10"	58	70	74				90AM	
Bowl, 4"	20	28	30					
Bowl, 7"	26	30	36				42AM	
Bowl, 9"	30	45	48				58AM	
OVAL PRISMS								
Hatpin		32						
OVAL STAR AND FAN (JENKINS)								
Rose Bowl	46	57						
OWL BANK								
One Size	38							
OWL BOTTLE								
One Shape							65CL	
OXFORD								
Mustard Pot, w/lid	50							
PAINTED CASTLE								
Shade	55							
PAINTED PANSY								
Fan Vase	42							
PALM BEACH (U.S. GLASS)								
Vase Whimsey	85	120					140W	
Bowl, 9"	50						75	
Bowl, 5"	30						50	
Butter	120						260AM	
Creamer, Spooner, Sugar ea	75						130AM	
Pitcher	450						680W	
Tumbler	100						165W	
Plate, 9", Rare	160	250					225	
Banana Bowl	100	220						
Rose Bowl Whimsey, Rare	85						210W	
PANAMA (U.S. GLASS)								
Goblet, Rare	120							
PANELLED CRUET								
One Size	95							
PANELLED DANDELION (FENTON)								
Pitcher	410	585	625	600				
Tumbler	52	58	56	70				
Candle Lamp Whimsey, Rare				3,000				
Vase Whimsey, Rare				2,700				
PANELLED DIAMOND AND BOWS (FENTON)								
Vase, 7"-14"	28	36	38	35	70		60	
Also called "Boggy Bayou"								
PANELLED HOBNAIL (DUGAN)								
Vase, 5"-10"	45	65	70		75		90	
PANELLED PALM (U.S. GLASS)								
Mug, Rare	90							
PANELLED PRISM								
Jam Jar, w/lid	48							
PANELLED SMOCKING								
Sugar	47							
PANELLED SWIRL								
Rose Bowl	65							
PANELLED THISTLE (HIGBEE)								
Tumbler	95							
PANELLED TREE TRUNK (DUGAN)								
Vase, 7"-12", Rare	70	95	110		150			
PANELS AND BALL (FENTON)								
Bowl, 11"	48						175W	
Also Called "Persian Pearl"								
PANELS AND BEADS								
Shade							44VO	
PANSY (IMPERIAL)								
Bowl, 8¾"	35	48	45				65AQ	
Creamer or Sugar	24	40	36				50SM	
Dresser Tray	56	90	85				110SM	
Pickle Dish, Oval	28	48	45				60SM	
Nappy, Old Only	18		24					
Plate, Ruffled, Rare	75	90	85				90SM	

Parlor Panels

Peach

Peacock and Dahlia

Peacock and Urn (Millersburg)

	M	A	G	B	PO	AO	Pas	Red
PANTHER (FENTON)								
Bowl, 10", Ftd	95	265	375	250			675W	
Bowl, 5", Ftd	55	90	95	80			675W	725
Whimsey Bowl, 10½"	800			900				
PAPERWEIGHT								
Flower-Shaped, Rare							195	
PARLOR								
Ashtray				95				
PARLOR PANELS								
Vase, 4"-11"	38	75					80SM	
PASTEL HAT								
Various Sizes	45						42+	
PASTEL PANELS (IMPERIAL)								
Pitcher							320	
Tumbler							70	
Mug, Stemmed							85	
Creamer or Sugar							60	
PEACH (NORTHWOOD)								
Bowl, 9"							210W	
Bowl, 5"							60W	
Butter							220W	
Creamer, Sugar, or Spooner, ea	260*						225W	
Pitcher				650			760W	
Tumbler				80			95W	
PEACH BLOSSOM								
Bowl, 7½"	55	70						
PEACH AND PEAR (DUGAN)								
Banana Bowl	70	95						
PEACHES								
Wine Bottle	37							
PEACOCK, FLUFFY (FENTON)								
Pitcher	500	700	750	850				
Tumbler	45	57	60	67				
PEACOCK (MILLERSBURG)								
Bowl, 9"	365	475	495				600CM	
Bowl, 5"	90	150	200	800				
Bowl, 7½", Vt, Rare	500	400	500					
Bowl, 6", Vt, Rare		145						
Ice Cream Bowl, 5"	60	80		140	450			
Plate, 6", Rare	900	775						
Spittoon Whimsey, Rare	4,500	7,000						
Proof Whimsey, Rare	250	245	260					
Rose Bowl Whimsey, Rare		3,000					3,200V	
Bowl, 10", Ice Cream, Rare	500	700	1,000					
Banana Bowl, Rare		3,000					4,500V	
PEACOCK AND DAHLIA (FENTON)								
Bowl, 7½"	50		125	95			150W	
Plate, 8½", Rare	300			300				
PEACOCK GARDEN (NORTHWOOD)								
Vase, 8", Rare	2,000						2,500	
PEACOCK AND GRAPE (FENTON)								
Bow, 7¾", Flat or Ftd	38	47	50	45	295		650IM	750
Plate 9" (Either Base), Rare	350	400		450			350	
PEACOCK LAMP								
Carnival Base		300	375					600
PEACOCK, STRUTTING (WESTMORELAND)								
Creamer or Sugar, w/lid		65	60					
PEACOCK TAIL (FENTON)								
Bon Bon	30	38	40	38				
Bowl, 4"-10"	28	32	36	32			225	
Compote	38	48	54	48				
Hat	32	38	46	38				
Hat, Advertising	40		50					
Plate, 9"	195	225		120				
Plate, 6"	50	70		70				
PEACOCK AND URN (FENTON)								
Bowl, 8½"	50	67	75	70	1,600			3,500
Plate, 9"	300	250	300	350			175W	
Compote	38	45	50	42			125V	795
Goblet, Rare	60	85		75			90V	

![Peacock at the Fountain (Northwood) - a footed bowl with ruffled edge](peacock_fountain)

Peacock at the Fountain
(Northwood)

Peacock Tail

People's Vase

Perfection

	M	A	G	B	PO	AO	Pas	Rec
PEACOCK AND URN (NORTHWOOD)								
Bowl, 9"	65	95	100					
Bowl, 10", Ice Cream	175	240	250	275		3,650	3,000AM	
Bowl, 5"	42	54	58					
Bowl, 6", Ice Cream	50	60	68	75		1,600	110IG	
Plate, 11", Rare	1,250	750						
Plate, 6", Rare	250	275						
(Add 10% If Stippled)								
PEACOCK AND URN AND VTS. (MILLERSBURG)								
Bowl, 9½"	185	245	260					
Compote, Large, Rare	900	950	995					
Bowl, 10", Ice Cream, Rare	275	350	425	1,800*				
Bowl, 6, Ice Cream, Rare	145	160	335	650				
6" Ruffled Bowl	60	70	95					
Plate, 10½", Rare	2,600	2,000						
Bowl, 8¼", Variant, Rare	350	450	550	1,100				
PEACOCK AT THE FOUNTAIN (DUGAN)								
Pitcher				395				
Tumbler		75		75				
PEACOCK AT THE FOUNTAIN (NORTHWOOD)								
Bowl, 9"	65	90	95	98			150	
Bowl, 5"	30	39	42	45			58	
Orange Bowl, Ftd	150	270	350	295		3,200	560	
Punch Bowl, w/base	450	575		550		10,000	2,000IB	
Cup	26	35		40			60	
Butter	225	375	395	310			560IB	
Sugar	100	190	195	210			235	
Creamer or Spooner	167	176	185	190			220	
Compote, Rare	295	325		345		3,500	1,100IB	
Pitcher	370		1,600				750IB	
Tumbler	36	48	475				85IB	
Spittoon Whimsey, Rare		3,200						
PEACOCK TAIL VT. (MILLERSBURG)								
Compote, Scarce	80	100	90					
PEACOCK TAIL AND DAISY								
Bowl, Very Rare	1,000	1,100*				1,200BO*		
PEACOCKS (ON FENCE) (NORTHWOOD)								
Bowl, 8¼"	80	110	125	135		750	700IG	
Plate, 9"	400	800	1,600	1,400		2,000	1,900IB	
PEARL AND JEWELS (FENTON)								
Basket, 4"							190W	
PEARL LADY (NORTHWOOD)								
Shade							55W	
PEARL #37 (NORTHWOOD)								
Shade							60W	
PEBBLE AND FAN (ENGLISH)								
Vase, 11¼", Rare				450			450AM	
PENNY								
Match Holder, Rare		245						
PEOPLE'S VASE (MILLERSBURG)								
Straight Top, Rare	8,000		9,000	9,000				
Ruffled Top, Rare		7,000						
PERFECTION (MILLERSBURG)								
Pitcher, Rare	4,000	4,000	4,500					
Tumbler, Rare	600	400	650					
PERSIAN GARDEN (DUGAN)								
Bowl, Ice Cream, 11"	270	375	385	390			310W	
Bowl, Ice Cream, 6"	60	70	80	90			90W	
Bowl, Berry, 10"	240	250					285	
Bowl, Berry, 5"	40	58			70		62	
Plate, Chop, 13", Rare		2,650			4,750		2,200W	
Plate, 6", Rare	90	140						
Fruit Bowl, w/base	210	370			495		320	

266

Persian Medallion

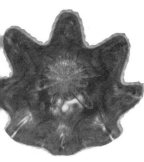

Petal and Fan

Pineapple

Pipe Humidor

	M	A	G	B	PO	AO	Pas	Red
PERSIAN MEDALLION (FENTON)								
Bon Bon	35	48	52	50			375IB	485
Compote	40	57	54	68				
Bowl, 10"	47	60	65	58				
Bowl, 5"	26	32	36	30				1,000
Bowl, 8¼"	42	50						795
Orange Bowl	85	185	195	175			200	
Rose Bowl	60	70	65	60			90	
Plate, 7"	70	80		75			95	
Plate, 9½"	125	148		250			500W	
Hair Receiver	60	70		65			95	
Punch Bowl/Base	250	350	390	325				
Punch Cup	25	30	35	28				
PETAL AND FAN (DUGAN)								
Bowl, 10"	48	68	70		195		120	
Bowl, 5"	30	44	42		50		60	
Bowl, 8½"	50	80						
Plate, 6", Ruffled		95			180			
PETALS (DUGAN)								
Bowl, 8¼"	40	52			110		62	
Compote	45	60					80	
Banana Bowl		90			110			
PETER RABBIT (FENTON)								
Bowl, 9", Rare	700		850	775				
Plate, 10", Rare	1,675		2,200	2,000			1,750AM	
PICKLE								
Paperweight, 4½"		45						
PIGEON								
Paperweight	80							
PILLAR AND DRAPE								
Shade					65MO		75W	
PILLOW AND SUNBURST (WESTMORELAND)								
Bowl, 7½"-9"	40	54			60		62AM	
PINE CONE (FENTON)								
Bowl, 6"	38	48	46	40			50AM	
Plate, 6½"	48	70	58	54				
Plate, 8", Rare			96	96				
PINEAPPLE (ENGLISH)								
Bowl, 7"	45	56		52				
Creamer	40							
Sugar, Stemmed or Flat	50							
Compote	50	65		52				
Butter	85							
PINEAPPLE, HEAVY (FENTON)								
Bowl, 10", Ftd, Rare	600*			800*			800*W	
PIN-UPS (AUSTRALIAN)								
Bowl, 8¼", Rare	90	110						
PINWHEEL (DUGAN)								
Bowl, 6"	38				57			
Plate, 6½"	58							
PINWHEEL (ENGLISH)								
Vase, 6½"	140	200						
Vase 8"	150							
Bowl, 8" Rare	165							
(Also Called Derby)								
PIPE HUMIDOR (MILLERSBURG)								
Tobacco Jar, w/lid, Rare	3,500	4,000	3,700					
PLAID (FENTON)								
Bowl, 8¼"	145		265	200			800IB	4,000
Plate, 9", Rare	325			385				
PLAIN JANE								
Paperwieght	80							
PLAIN JANE (IMPERIAL)								
Basket	60							
PLAIN PETALS (NORTHWOOD)								
Nappy, Scarce		85	90					
(Interior of Leaf and Beads Nappy)								
PLEATS AND HEARTS								
Shade							70	

267

Poinsettia (Imperial)

Poinsetta (Northwood)

Poppy

Poppy Show Vase

	M	A	G	B	PO	AO	Pas	Re
PLUME PANELS								
Vase, 7"-12"	40	50	52	45			165	490
POINSETTIA (IMPERIAL)								
Milk Pitcher	185	850	250				230SM	
POINSETTIA (NORTHWOOD)								
Bowl, 8½", Flat or Ftd	160	275	480	290		1,050	400	
POLO								
Ashtray	40							
POND LILY (FENTON)								
Bon Bon	38		50	52			68W	
PONY (DUGAN)								
Bowl, 8½"	70	120					470AQ	
Plate, 9", Rare, Age Questionable	450*							
POODLE								
Powder Jar, w/lid	25							
POPPY (MILLERSBURG)								
Compote, Scarce	650	550	400					
Salver, Rare	1,100	1,300	1,450					
POPPY AND FISH NET (IMPERIAL)								
Vase, 6", Rare								500
POPPY SHOW (IMPERIAL)								
Vase, 12", Old Only	450	1,500	850				900SM	
Hurricane Whimsey		2,000					2,000W	
POPPY SHOW (NORTHWOOD)								
Bowl, 8½"	320	400	460	400			1,850IB	
Plate, 9" Rare	510	700	2,000	630		20,000	875W	
POPPY WREATH (NORTHWOOD)								
(Amaryllis Exterior Pattern)								
PORTLAND (U.S. GLASS)								
Bowl, 8½"							150	
POTPOURRI (MILLERSBURG)								
Milk Pitcher, Rare	1,600							
PRAYER RUG (FENTON)								
Bon Bon, Rare					1,400IC			
(Iridized Custard Only)								
PREMIUM (IMPERIAL)								
Candlesticks, pr	60	95						110
Bowl, 8½"	45	80						95
Bowl, 12"	60	90						100
Under Plate, 14"	65	110						130
PRETTY PANELS (FENTON)								
Pitcher, w/lid								490
Tumbler, Handled	52							70
PRETTY PANELS (NORTHWOOD)								
Pitcher	120		160					
Tumbler	60		70					
PRIMROSE (MILLERSBURG)								
Bowl, 8¼", Ruffled	90	120	150	4,000			150CM	
Bowl, 9", Ice Cream, Scarce	110	140	150					
Bowl, Experimental, Rare, Goofus Exterior		800						
PRIMROSE AND FISHNET (IMPERIAL)								
Vase, 6", Rare								500
PRIMROSE PANELS (IMPERIAL)								
Shade							45	
PRINCELY PLUMES								
Candle Holder		260*						
PRINCESS (U.S. GLASS)								
Lamp, Complete, Rare		1,250						
PRISM								
Shakers, pr	60							
Tray, 3"	45							
Hatpin		40						

Propeller

Puzzle

Rambler Rose

Ranger

	M	A	G	B	PO	AO	Pas	Red
PRISM BAND (FENTON)								
Pitcher, Decorated	165	345	375	385			325W	
Tumbler, Decorated	26	45	48	38			150W	
PRISM AND CANE (ENGLISH)								
Bowl, 5", Rare	42	65						
PRISM AND DAISY BAND (IMPERIAL)								
Vase	28							
Compote	35							
Bowl, 5"	18							
Bowl, 8"	30							
Sugar or Creamer, ea	35							
PRISMS (WESTMORELAND)								
Compote, 5", Scarce	80	110	125				200AQ	
PROPELLER (IMPERIAL)								
Bowl, 9½", Rare	80							
Compote	30		38					
Vase, Stemmed, Rare	75							
PROUD PUSS (CAMBRIDGE)								
Bottle	80							
PULLED LOOP (DUGAN)								
Vase	28	36	35	30	50			
PUMP, HOBNAIL (NORTHWOOD)								
One Shape, Age Questionable			850					
PUMP, TOWN (NORTHWOOD)								
One Shape, Rare	1,000	650	1,250					
PUZZLE (DUGAN)								
Compote	36	46	50	45	75		60W	
Bon Bon, Stemmed	38	48	54	48	62		65W	
PUZZLE PIECE								
One Shape				100				
QUARTERED BLOCK								
Creamer	50							
Sugar	50							
Butter	90							
QUEEN'S LAMP								
One Shape, Rare		2,000						
QUESTION MARKS (DUGAN)								
Bon Bon	38	48			70		75IG	
Compote	40	52			75		65W	
Cake Plate, Stemmed, Rare		450*						
QUILL (DUGAN)								
Pitcher, Rare	1,400	2,700						
Tumbler, Rare	375	450						
RAGGED ROBIN (FENTON)								
Bowl, 8¼", Scarce	70	100	95	80			150W	
RAINBOW (NORTHWOOD)								
Compote		110	145					
RAINDROPS (DUGAN)								
Bowl, 9"	50	60			58			
Banana Bowl, 9¼"		125			150			
RAMBLER ROSE (DUGAN)								
Pitcher	150	245	260	200				
Tumbler	30	35	45	40				
RANGER (MEXICAN)								
Creamer	40							
Nappy	80							
Tumbler	270							
Milk Pitcher	165							
Sugar	140							
Butter	90							
Breakfast Set, 2 Pieces	160							
Pitcher, Rare	285							
RANGER TOOTHPICK								
Toothpick Holder	90							

Rays and Ribbons

Ripple

Rising Sun

Rococo

	M	A	G	B	PO	AO	Pas	R
RASPBERRY (NORTHWOOD)								
Bowl, 9"	45	56	62					
Bowl, 5"	28	35	37					
Milk Pitcher	120	150	160				2,000IB	
Sauce Boat, Ftd	80	95		160				
Pitcher	140	220	240				1,600W	
Tumbler	36	42	46				350IG	
Compote	48	56	58					
RAYS (DUGAN)								
Bowl, 5"	40	48	48		75			
Bowl, 9"	55	90	90		125			
RAYS AND RIBBONS (MILLERSBURG)								
Bowl, 8½"-9½", Round or Ruffled	60	80	90				300V	
Plate, Rare	1,100							
Banana Bowl, Rare			900					
Bowl, Tri-Cornered or Square	115	125	135					
RED PANELS (IMPERIAL)								
Shade								20
REGAL IRIS (CONSOLIDATED GLASS)								
Gone-With-The-Wind Lamp, Rare	3,000							9,70
REGAL SWIRL								
Candlestick, ea	70							
REX								
Pitcher	365							
Tumbler	55							
RIB AND PANEL (FENTON)								
Vase	45							
Spittoon Whimsey	90							
RIBBED ELIPSE								
Mug, Rare							90CM	
RIBBED SWIRL								
Tumbler	56		70					
RIBBON AND BLOCK								
Lamp, Complete	500							
RIBBON AND FERN								
Atomizer, 7"	75							
RIBBON AND LEAVES								
Sugar, Small	48							
RIBBON TIE (FENTON)								
Bowl, 8¾"	46	60	60	55				1,60
Plate, Ruffled, 9"				140				2,00
Plate, Flat, 9½"				285				
RINGS								
Vase, 8"	55							
RIPPLE (IMPERIAL)								
Vase, Various Sizes	30	45	40	40			50	
RISING SUN (U.S. GLASS)								
Butterdish	150							
Creamer	75							
Sugar	85							
Pitcher, Rare, 2 Shapes	950			1,850				
Tumbler, Rare	400			750				
Tray, Rare				500				
ROBIN (IMPERIAL)								
Mug, Old Only	55						150SM	
Pitcher, Old Only, Scarce	350							
Tumbler, Old Only, Scarce	60							
ROCK CRYSTAL (McKEE)								
Punch Bowl, w/base		590						
Cup		45						
ROCOCO (IMPERIAL)								
Bowl, 5"	30		140				100SM	
Vase, 5½"	65		160				150SM	
ROLL								
Tumbler	38							
Cordial Set, Complete (Decanter, Stopper, 6 Glasses)	250							
Pitcher, Rare							275CL	
Shakers, ea, Rare	40							
ROMAN ROSETTE (U.S. GLASS)								
Goblet, 6", Rare							90CL	
ROOD'S CHOCOLATES (NORTHWOOD)								
Advertising Plate		245						

Rose Column

Rose Garden

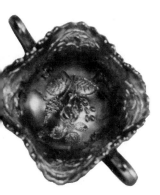

Roses and Fruit

Roses and Ruffles

	M	A	G	B	PO	AO	Pas	Red
ROSALIND (MILLERSBURG)								
Bowl, 10" Scarce	185	260	275				600AQ	
Bowl, 5", Rare			550					
Compote, 6", Rare (Variant)		375	375					
Compote, 8", Ruffled, Rare	850							
Compote, 9", Jelly, Rare	1,600	1,800	1,800					
ROSE								
Bottle							120	
ROSE BOUQUET								
Creamer	54							
ROSE COLUMN (MILLERSBURG)								
Vase, Rare	1,500	1,300	1,150	8,000				
Experimental Vase, Rare		4,000						
ROSE GARDEN (SWEDEN)								
Letter Vase	40			55				
Bowl, 8¼"	55	65		60				
Vase, 9" Round	200							
Pitcher, Communion, Rare	1,200			1,600				
Bowl, 6", Rare		75						
Butter, Rare	120			145				
Rose Bowl, Rare		175						
ROSE AND GREEK KEY								
Square Plate, Rare							6,000AM	
ROSE PANELS (AUSTRALIAN)								
Compote, Large	120							
ROSE PINWHEEL								
Bowl, Rare	1,700		2,000					
ROSE SHOW (NORTHWOOD)								
Bowl, 8¼"	237	300	295	295	2,500IC	900	350	
ROSE SHOW VARIANT (NORTHWOOD)								
Plate, 9"	300	340	355	650	875	1,000	4,500IC	
Bowl, 8¼"	200	225		350			250W	
ROSE SPRAY (FENTON)								
Compote	160						180	
ROSE TREE (FENTON)								
Bowl, 10" Rare	350			400				
ROSE WREATH (Basket of Roses) (NORTHWOOD)								
Bon Bon, Rare	200*	250*		250*				
(Basketweave Exterior)								
ROSES AND FRUIT (MILLERSBURG)								
Bon Bon, Ftd, Rare	565	700	700	1,000*				
ROSES AND RUFFLES (CONSOLIDATED GLASS)								
G-W-T-W Lamp, Rare	1,850							6,000
ROSETIME								
Vase	85							
ROSETTES (NORTHWOOD)								
Bowl, 9", Dome Base	58	85						
Bowl, 7", Ftd		95						
ROUND-UP (DUGAN)								
Bowl, 8¼"	54	75		80	97		140W	
Plate, 9", Rare	128	160		175	360		210V	
ROYALTY (IMPERIAL)								
Punch Bowl, w/base	125							
Cup	28							
Fruit Bowl, w/stand	100						135SM	
RUFFLED RIB (NORTHWOOD)								
Spittoon Whimsey, Rare	200							
Bowl, 8"-10"	46	60						
Vase, 7"-14"	60							
RUFFLES AND RINGS (NORTHWOOD)								
Bowl, Very Rare					900			
RUFFLES, RINGS AND DIASY BAND (NORTHWOOD)								
Bowl, 8½", Ftd, Rare		90						

S-Repeat

Sailboats

Scotch Thistle

Scroll and Flower Panels

	M	A	G	B	PO	AO	Pas	Red
RUSTIC								
(FENTON)								
Vase, Funeral, 15"-20"	140	200	220	185				
Vase, Various Sizes	28	32	35	32	75		60	550
S-BAND								
(AUSTRALIAN)								
Compote	50	65						
S-REPEAT								
(DUGAN)								
Creamer, Small		60						
Punch Bowl, w/base, Rare		1,700						
Cup, Rare		110						
Toothpick Holder (Old Only),								
Rare		60						
Tumbler	45							
Sugar, Rare (Very Light								
Iridescence)		185						
SACIC (ENGLISH)								
Ashtray	75							
SAILBOATS								
(FENTON)								
Bowl, 6"	28		75	60			200	350
Goblet	230	350	260	60			175	
Wine	30			100				
Compote	37			150				
Plate	450			425			295	
SAILING SHIP								
PLATE, 8"	40							
SAINT								
(ENGLISH)								
Candlestick, ea	275							
SALAMANDERS								
Hatpin		45						
SALT CUP								
One Shape	45						55CeB	
SATIN SWIRL								
Atomizer							65CL	
SCALE BAND (FENTON)								
Bowl, 6"	26				40			
Plate, Flat, 6½"	40						68V	410
Plate, Dome Base, 7"	45							400
Pitcher	110			210				
Tumbler	26			35				
SCALES								
(WESTMORELAND)								
Bon Bon	40	48			90	300	60TL	
Bowl, 7"-10"	32	52			70		46TL	
Deep Bowl, 5"		40						
Plate, 6"	45	58					65TL	
Plate, 9"		95			110	260	140	
SCARAB								
Hatpin		35						
SCOTCH THISTLE (FENTON)								
Compote	46	60	75	45				
SCOTTIE								
Powder Jar, w/lid	25							
Paperweight, Rare	200							
SCROLL (WESTMORELAND)								
Pin Tray	45							
SCROLL AND FLOWER PANELS								
(IMPERIAL)								
Vase, 10", Old Only	90	200		260				
SCROLL EMBOSSED								
(IMPERIAL)								
Bowl, 8½"	36	45					55AQ	485
Dessert, Round or Ruffled,								
Rare	90	110	120					
Plate, 9"	58	97						
Compote, Large	50	65						
Compote, Small	38	48	60				65AQ	
SCROLL EMBOSSED VT.								
(ENGLISH)								
Handled Ashtray, 5"	45	60						
Plate 7"	150							
SEACOAST								
(MILLERSBURG)								
Pin Tray, Rare	450	475	300					
SEAFOAM								
(DUGAN)								
Exterior Pattern Only								
SEAGULL								
Vase, Rare	750							

	M	A	G	B	PO	AO	Pas	Red
SEAGULLS								
Bowl, 6½", Scarce	80							
SEAWEED								
Lamp, 2 Sizes	195							
Lamp Vt. 8½", Rare							350IB	
SEAWEED (MILLERSBURG)								
Bowl, 5", Rare	400		450					
Bowl, 9", Rare	265		350	1,100			385AM	
Plate, 10", Rare	800	900	900					
Bowl, 10½", Ruffled, Scarce	185	375	295				200CM	
Bowl, 10½", Ice Cream, Rare	400	450	450					
SERRATED RIBS								
Shaker, ea	50							
SHARP								
Shot Glass	40						60SM	
SHELL								
Shade							65	
SHELL (IMPERIAL)								
Bowl, 7"-9"	38	46	50				85SM	
Plat, 8½"	80	145						
SHELL AND BALLS								
Perfume, 2½"	48							
SHELL AND JEWEL (WESTMORELAND)								
Creamer, w/lid	55	65	60				90W	
Sugar, w/lid	55	65	60				90W	
SHERATON (U.S. GLASS)								
Pitcher							165*	
Tumbler							48*	
Butter							120*	
Sugar							80*	
Creamer or Spooner							65*	
SHIP AND STARS								
Plate, 8"	30							
SHRINE (U.S. GLASS)								
Champagne							160CL	
Toothpick Holder		515					175CL	
SIGNET (ENGLISH)								
Sugar, w/lid, 6½"	70							
SILVER AND GOLD								
Pitcher	100							
Tumbler	25							
SILVER QUEEN (FENTON)								
Pitcher	175							
Tumbler	45							
SINGING BIRDS (NORTHWOOD)								
Bowl, 10"	52	60	75					
Bowl, 5	30	35	42					
Mug	195	120	350	280		1,800	1,250IB	
Butter	185	295	320					
Sugar	110	140	155					
Creamer	80	100	120					
Spooner	80	100	120					
Pitcher	325	365	390					
Tumbler	45	54	58					
Sherbet, Rare		85						
SINGLE FLOWER (DUGAN)								
Bowl, 8"	26	34	38		70			
Hat	24	32	35					
Handled Basket Whimsey, Rare	225			850				
Banana Bowl, 9½", Rare				325				
SINGLE FLOWER FRAMED (DUGAN)								
Bowl, 8¾"	55	90	90		125			
Bowl 5"	40	48			75			
SIX PETALS (DUGAN)								
Bowl, 8½"	38	45	48	52	90		60	
Plate, Rare	82	120	145		190			
Hat	40	45	50		70		125BA	
SIX-SIDED (IMPERIAL)								
Candlestick, ea	150	275	275				200SM	

Seaweed

Shell and Jewel

Silver Queen

Six-Sided

273

Ski Star

Smooth Panels

Snow Fancy

Soutache

	M	A	G	B	PO	AO	Pas	Re
SKATER'S SHOE								
(U.S. GLASS)								
One Shape	90							
SKI-STAR (DUGAN)								
Bowl, 8"-10"	58	90		165	95			
Bowl, 5"	32	47	50	50				
Basket, Handled, Rare					500			
Banana Bowl		120			250			
Hand Grip Bowl, 8"-10"		90						
Rose Bowl, Rare					500			
SMALL BASKET								
One Shape	48							
SMALL BLACKBERRY								
(NORTHWOOD)								
Compote	45	57	57					
SMALL PALMS								
Shade	34							
SMALL RIB (DUGAN)								
Compote	35	40	40				45AM	
Rose Bowl, Stemmed	38	42	42				48AM	
SMALL THUMBPRINT								
Creamer	60							
Toothpick Holder	50							
SMOOTH PANELS								
(IMPERIAL)								
Bowl, 6½"							35PM	
Plate, 9¼"							85CL	
Tumbler							42	
Vase	36	50	50		65		58SM	
Pitcher	85		170					
SMOOTH RAYS								
(IMPERIAL)								
Bon Bon	26						24CL	
SMOOTH RAYS								
(NORTHWOOD)								
Bon Bon			50					
SMOOTH RAYS								
(NORTHWOOD-DUGAN)								
Compote	36	50	52					
Rose Bowl	40	55						
Bowl, 6"-9"	44	48			60			
Plate, 7"-9"	60				80			
SMOOTH RAYS								
(WESTMORELAND)								
Compote			70				65AM	
Bowl, 7"-9", Flat	38	50	47		70		75TL	
Bowl, Dome Base, 5"-7½"			50		70		75TL	
SNOW FANCY (McKEE)								
Bowl, 5"			40					
Creamer or Sugar	47							
SODA GOLD								
(IMPERIAL)								
Candlestick, 3½", ea	32						40	
Bowl, 9"	45						57	
Pitcher	210						325	
Tumbler	40						75	
SODA GOLD SPEARS								
(DUGAN)								
Bowl, 8½"	38						40CL	
Bowl, 4½"	26						27CL	
Plate, 9"	50						160CL	
SOLDIERS AND SAILORS								
(FENTON)								
Plate (Illinois) Rare	900	900		1,000				
Plate (Indianapolis), Rare				3,000				
SOUTACHE								
(DUGAN)								
Bowl, 10"					185			
Plate, 10½", Rare					365			
Lamp, Complete	285							
SOUTHERN IVY								
Wine, 2 Sizes	38							
SOUVENIR BANDED								
Mug	70							
SOUVENIR BELL								
(IMPERIAL)								
One Shape, Lettering	170							
SOUVENIR MINIATURE								
One Shape, Lettering	45							
SOUVENIR MUG (McKEE)								
Any Lettering	52							

Spiralex

Split Diamond

Springtime

Star

	M	A	G	B	PO	AO	Pas	Red
SOUVENIR PIN TRAY (U.S. GLASS)								
One Size							75	
(Same as Portland Pattern)								
SOUVENIR VASE (U.S. GLASS)								
Vase, 6½", Rare...................	90	120			150	350		
SOWERBY FLOWER BLOCK (ENGLISH)								
Flower Frog	60							
SPHINX (ENGLISH)								
Paperweight, Rare...............							475AM	
SPIDERWEB (NORTHWOOD)								
Candy Dish, covered............							35SM	
SPIDERWEB (NORTHWOOD-DUGAN)								
Vase, 8"	45						75	
SPIDERWEB AND TREEBARK (DUGAN)								
Vase, 6"							60	
SPIRAL (IMPERIAL)								
Candlestick, ea...................	55	70	75				90	
SPIRALEX (ENGLISH)								
Vase, Various Sizes..............	45	58	58	50			60	
SPIRALLED DIAMOND POINT								
Vase, 6"	38							
SPLIT DIAMOND (ENGLISH)								
Creamer, Small...................	36							
Sugar, Open	40							
Butter, Scarce	70							
SPOKES (FOSTORIA)								
Bowl 10"............................							95	
SPRING BASKET (IMPERIAL)								
Handled Basket, 5"..............	40						48SM	
SPRING OPENING (MILLERSBURG)								
Plate, 6½", Rare		400						
SPRINGTIME (NORTHWOOD)								
Bowl, 9".............................	100	200	220					
Bowl, 5".............................	40	50	56					
Butter	370	425	435					
Sugar	335	400	415					
Creamer or Spooner	320	390	400					
Pitcher, Rare......................	750	1,000	1,200				1,800	
Tumbler, Rare	100	150	185				300	
SQUARE DAISY AND BUTTON (IMPERIAL)								
Toothpick Holder, Rare							120	
SQUARE DIAMOND								
Vase, Rare			125					
STAG AND HOLLY (FENTON)								
Bowl, 9"-13", Ftd	95	275	290	250	1,700		580AQ	2,400
Rose Bowl, Ftd	385		800	1,000				
Plate, 13", Ftd.....................	1,000							
Plate, 9", Ftd.......................	700	900		1,800				
STANDARD								
Vase, 5½"............................	48							
STAR								
Paperweight, Rare...............							265	
STAR (ENGLISH)								
Bowl, 8"..............................	40							
STAR CENTER (IMPERISL)								
Bowl, 8½"	30	36					42	
Plate, 9".............................	60	80					90	
STAR OF DAVID (IMPERIAL)								
Bowl, 8¾", Scarce	60	80	80				100SM	3,000*
STAR OF DAVID AND BOWS (NORTHWOOD)								
Bowl, 8½"	45	60	70				140AM	
STAR AND DIAMOND POINT								
Hatpin................................		50						
STAR AND FAN								
Vase, 9½", Rare...................	250			200				
(Note: Same as curved Star Pattern)								
STAR AND FAN (ENGLISH)								
Cordial Set	1,500							
(Decanter, 4 Stemmed Cordials and Tray)								

	M	A	G	B	PO	AO	Pas	Re

Star and File

STAR AND FILE (IMPERIAL)

	M	A	G	B	PO	AO	Pas	Re
Bowl, 7"-9½"	30						40	
Vase, Handled	50						40	
Compote	45						56	
Creamer or Sugar	28							
Pitcher	185							
Tumbler	140							
Decanter, w/stopper	110							
Rose Bowl	65	90					90AM	
Wine	40						200IG	
Spooner	30							
Sherbet	32							
Custard Cup	26							
Plate, 6"	65							
Pickle Dish	40							
Tumbler, Rare	265							
Bon Bon	30							

STAR AND HOBS

	M	A	G	B	PO	AO	Pas	Re
Rose Bowl, 9", Rare				200				

STAR MEDALLION (IMPERIAL)

	M	A	G	B	PO	AO	Pas	Re
Bon Bon	45		60					
Bowl, 7"-9	28						38	
Compote	45							
Butter	100							
Creamer, Spooner, or Sugar ea	60							
Milk Pitcher	80		75					
Goblet	45						60	
Tumbler	30		45				52	
Plate, 5"	60						45	
Plate, 10"							50CL	
Handled Celery	80						65	
Celery Tray	60						50	
Ice Cream, Stemmed, Small	35							
Pickle Dish	40						45	
Vase, 6"	40							
Custard Cup	20							

Star Medallion

STAR AND NEARCUT

	M	A	G	B	PO	AO	Pas	Re
Hatpin		38						

STAR AND ROSETTE

	M	A	G	B	PO	AO	Pas	Re
Hatpin		38						

STAR SPRAY (IMPERIAL)

	M	A	G	B	PO	AO	Pas	Re
Bowl, 7"	28						30	
Bride's Basket, Complete, Rare	80						95	
Plate, 7½", Scarce	50						75	

STARBRIGHT

	M	A	G	B	PO	AO	Pas	Re
Vase, 6½"	37	42		42				

STARBURST

	M	A	G	B	PO	AO	Pas	Re
Perfume, w/stopper	50							

STARFISH (DUGAN)

	M	A	G	B	PO	AO	Pas	Re
Bon Bon, Handled, Rare					145		800	
Compote	40	60	60		90			

STARFLOWER

	M	A	G	B	PO	AO	Pas	Re
Pitcher, Rare	2,600		2,000					

STARLYTE (IMPERIAL)

	M	A	G	B	PO	AO	Pas	Re
Shade	28							

Stippled Diamond Swag

STARS AND BARS

	M	A	G	B	PO	AO	Pas	Re
Wine	40							

STARS AND STRIPES (OLD GLORY)

	M	A	G	B	PO	AO	Pas	Re
Plate, 7½", Rare	125							

STIPPLED ACORNS

	M	A	G	B	PO	AO	Pas	Re
Candy dish, Ftd, w/lid	75	90		90				

STIPPLED DIAMOND SWAG (ENGLISH)

	M	A	G	B	PO	AO	Pas	Re
Compote	45		56*	56*				

STIPPLED FLOWER (DUGAN)

	M	A	G	B	PO	AO	Pas	Re
Bowl, 8½"					80			

STIPPLED PETALS (DUGAN)

	M	A	G	B	PO	AO	Pas	Re
Bowl, 9"		80			72			
Handled Basket		170			160			
Bowl, 9"					90			

STIPPLED RAMBLER ROSE (DUGAN)

	M	A	G	B	PO	AO	Pas	Re
Nut Bowl, Ftd	60			75				

Stippled Rambler Rose

Stippled Strawberry

Strawberry Intaglio

Strawberry Scroll

Studs

	M	A	G	B	PO	AO	Pas	Red
STIPPLED RAYS (FENTON)								
Bon Bon	28	35	36	35				350
Bowl, 5"-9"	30	37	35	35				300
Compote	34	40	42	40				
Creamer or Sugar, ea	25	32	35	35				400
Plate, 7"	30	42	45	40				450
STIPPLED RAYS (IMPERIAL)								
Creamer, Stemmed	40		46				50SM	
Sugar, Stemmed	40		46				50SM	
STIPPLED RAYS (NORTHWOOD)								
Bowl, 8"-10"	46	55	60					
Compote	45	58	60					
STIPPLED SALT CUP								
One Size	40							
STIPPLED STRAWBERRY (JENKINS)								
Tumbler	75							
Creamer or Sugar	30							
Spittoon Whimsey, Rare	220							
Syrup, Rare	200							
Butter	90							
STORK (JENKINS)								
Vase	50							
STORK ABC								
Child's Plate, 7½"	75							
STORK AND RUSHES (DUGAN)								
Butter, Rare	145	165						
Creamer or Spooner, Rare	80	100						
Sugar, Rare	100	120						
Bowl, 10"	45	50						
Bowl, 5"	25	29						
Mug	30	58		350				
Punch Bowl, w/base, Rare	200	315		340				
Cup	18	35		30				
Hat	20			28				
Handled Basket	50							
Pitcher	250			450				
Tumbler	30	60		70				
STRAWBERRY (DUGAN)								
Epergne, Rare	1,100	900						
STRAWBERRY (FENTON)								
Bon Bon	30	45	65	60			110V	450
STRAWBERRY (MILLERSBURG)								
Bowl, 6½"	90	140	150					
Bowl, 8"-10", Scarce	250	300	300				300CM	
Compote, Rare	150	200	220				850V	
Gravy Boat Whimsey, Rare							400V	
Banana Boat Whimsey, Rare		2,000	2,000				1,900V	
Bowl, 9½", Tri-Cornered	350	600	400					
STRAWBERRY (NORTHWOOD)								
Bowl, 8"-10"	45	56	54	50		1,850	750IG	
Bowl, 5"	30	40	40	40				
Plate, 9"	175	240	250	200		500		
Plate, Handgrip, 7"	175	100	200					
STRAWBERRY INTAGLIO (NORTHWOOD)								
Bowl, 9½"	50							
Bowl, 5½"	25							
STRAWBERRY POINT								
Tumbler	100							
STRAWBERRY SCROLL (FENTON)								
Pitcher, Rare	1,800			2,000				
Tumbler, Rare	150			175				
STRAWBERRY SPRAY								
Brooch				170				
STRETCHED DIAMOND (NORTHWOOD)								
Tumbler, Rare	290							
STREAM OF HEARTS (FENTON)								
Bowl, 10", Ftd	70			95				
Compote, Rare	90						120	
STRING OF BEADS								
Shape	30		36					
STUDS (IMPERIAL)								
Tray, Large	60							
Juice, Tumbler	30							
Milk, Pitcher	70							

Sunflower

Sunflower and Diamond

Superb Drape

Sweetheart

	M	A	G	B	PO	AO	Pas	Re
STYLE								
Bowl, 8"..........................		92						
SUMMER DAYS (DUGAN)								
Vase, 6".............................	50			60				
(Note: This is actually the base for the Stork and Rushes punch set)								
SUN PUNCH								
Bottle	24						26	
SUNFLOWER (MILLERSBURG)								
Pin Tray, Rare	425	350	300					
SUNFLOWER (NORTHWOOD)								
Bowl, 8½"	45	65	58				80	
Plate, Rare..........................	150		325					
SUNFLOWER AND DIAMOND								
Vase, 2 Sizes.......................	65			90				
SUNGOLD (AUSTRALIAN)								
Epergne..............................							250	
SUNK DIAMOND BAND (U.S. GLASS)								
Pitcher, Rare......................	145						215W	
Tumbler, Rare	50						70W	
SUNKEN DAISY (ENGLISH)								
Sugar	28		36					
SUNKEN HOLLYHOCK								
G-W-T-W Lamp, Rare	3,000							9,900
SUNRAY								
Compote.............................		38			45			
SUNRAY (FENTON)								
Compote (Milk Glass Iridized)					90MO			
SUPERB DRAPE (NORTHWOOD)								
Vase, Rare						2,500		
SWAN, PASTEL (DUGAN-FENTON)								
One Size	75	280		450			35IB	
SWEETHEART (CAMBRIDGE)								
Cookie Jar, w/lid, Rare........	1,300		750					
Tumbler, Rare	600							
SWIRL (NORTHWOOD)								
Pitcher...............................	180		225					
Tumbler	48		65					
Mug, Rare	75							
Candlestick, ea...................	30							
SWIRL HOBNAIL (MILLERSBURG)								
Rose Bowl, Rare	275	350	585					
Spittoon, Rare	550	700	1,050					
Vase, 7"-10", Rare...............	225	275	250	400				
SWIRL VARIANT (IMPERIAL)								
Bowl, 7"-8"	28				45		38	
Epergne..............................			170		200			
Vase, 6½"............................	26		40				50W	
Plate, 6"-8¾"	45		50				60	
Pitcher, 7½"	145							
Cake Plate							75CL	
Dessert, Stemmed	28							
Juice Glass.........................	30							
SWIRLED FLUTE (FENTON)								
Vase, 7"-12"........................	28	34	40	32			60W	475
SWIRLED RIBS (NORTHWOOD)								
Pitcher...............................	165							
Tumbler	60	70						
SWORD AND CIRCLE								
Tumbler, Rare	85							
SYDNEY (FOSTORIA)								
Tumbler, Rare	400							
TAFFETA LUSTRE (FOSTORIA)								
Candlestick, pr, Rare		300	300	300			400AM	
Console Bowl, 11", Rare.......		150	150	150			165AM	
(Add 25% For Old Paper Labels Attached)								
Perfume, w/stopper............	65	75						
TALL HAT								
Various Sizes, 4"-10"	40						50PK	
TARGET (FENTON)								
Vase, 7"-11"........................	32	46	48		85		48	
TEXAS								
Giant Tumbler.....................				185				

Thistle (English)

Thistle and Thorn

Thistle Banana Boat

Tiger Lily

	M	A	G	B	PO	AO	Pas	Red
TEXAS HEADDRESS (WESTMORELAND)								
Punch Cup	40							
TEN MUMS (FENTON)								
Bowl, 9", Ftd, Rare	400							
Bowl, 8"-11"	90	120	110	100				
Plate, 10", Rare...................			375	380				
Pitcher, Rare......................	425			875			1,400W	
Tumbler, Scarce	70			80			280W	
THIN RIB (FENTON)								
Candlestick, pr	60							375
THIN RIB AND DRAPE (FENTON)								
Vase, 8"-14"	36	48	48					
THIN RIB (NORTHWOOD) AND VTS								
Vase, 6"-11"	26	35	45	42	56	190	50	
THISTLE (ENGLISH)								
Vase, 6"	30							
THISTLE (FENTON)								
Bowl, 8"-10"	100	150	160	150			200AQ	
Plate, 9", Rare......................		1,400	3,000					
Compote...........................	55			58				
Advertising Bowl (Horlacher)...	110	180	175	170				
THISTLE								
Shade................................	40							
THISTLE AND LOTUS (FENTON)								
Bowl, 7"............................	50		65	60				
THISTLE AND THORN (ENGLISH)								
Bowl, 6", Ftd.......................	46							
Creamer or Sugar, ea...........	50							
Plate, 8½", Ftd	130							
Nut Bowl	70							
THISTLE, FENTON'S (FENTON)								
Banana Boat, Ftd, Scarce	325		460	400				
THREE DIAMONDS (DUGAN)								
Vase, 6"-10"	30	46	48	45	75		50	
THREE FLOWERS (IMPERIAL)								
Tray, Center Handle, 12"	47						52SM	
THREE FRUITS (NORTHWOOD)								
Bowl, 9".............................	36	50	54	56		1,000	90	
Bowl, 5".............................	22	28	30	35		200	55	
Bowl, 8⅝", Dome Base		75						
Bon Bon, Stemmed..............	46	58	60	65		700	95	
Plate, 9", Round..................	75	100	110	135		3,900	700IB	
(Add 10% if Stippled)								
THREE FRUITS MEDALLION (NORTHWOOD)								
Bowl, 10½", Ftd, Rare (Meander Exterior)...............	110	155	170	275		1,200	750BA	
THREE FRUITS VT. (DUGAN)								
Plate, 12-Sided	150	200	200				135	
THREE-IN-ONE (IMPERIAL)								
Bowl, 8¾"...........................	28	36	36				42SM	
Bowl, 4½"	18	22	22				26SM	
Plate, 6½"	55						80SM	
Rose Bowl, Rare	195							
Banana Bowl Whimsey	100							
Toothpick Holder, Rare			90					
THREE MONKEYS								
Bottle, Rare							75	
THREE ROW (IMPERIAL)								
Vase, Rare	800	1,000						
THUMBPRINT AND OVAL (IMPERIAL)								
Vase, 5½", Rare....................	400	600						
THUMBPRINT AND SPEARS								
Creamer	50		56					
THUNDERBIRD (AUSTRALIAN)								
Bowl, 9½"	167	185						
Bowl, 5".............................	35	45						
TIGER LILY (IMPERIAL)								
Pitcher..............................	130	350	300					
Tumbler	32	55	38				90	
TINY BERRY								
Tumbler, 2¼"				40				

279

	M	A	G	B	PO	AO	Pas	Re
TINY HOBNAIL								
Lamp	95							
TOBACCO LEAF								
(U.S. GLASS)								
Champagne							160CL	
TOLTEC (McKEE)								
Butter (Ruby Iridized), Rare		350						
Pitcher, Tankard, Very Rare	2,000							
TOMOHAWK (CAMBRIDGE)								
One Size, Rare				1,650				
TOP HAT								
Vase							45	
TOP O' THE MORNING								
Hatpin		35						
TOP O' THE WALK								
Hatpin		60		65				
TORNADO (NORTHWOOD)								
Vase, Plain	420	450	470	900			1,000W	
Vase, Ribbed, 2 Sizes	450	470	495	890			1,300IB	
Vase Whimsey							450WS	
TORNADO VT.								
(NORTHWOOD)								
Vase, Rare	1,200*							
TOWERS								
(ENGLISH)								
Hat Vase	40							
TOY PUNCH SET								
(CAMBRIDGE)								
Bowl Only, Ftd	48							
TRACERY								
(MILLERSBURG)								
Bon Bon, Rare		650	600					
TREE BARK								
(IMPERIAL)								
Pitcher, Open Top	60							
Pitcher, w/lid	70							
Tumbler, 2 Sizes	24							
Bowl, 7½"	18							
Pickle Jar, 7½"	35							
Candlestick, 7", pr	60							
Sauce, 4"	10							
Candlestick, 4½", pr	30							
Candy Jar, w/lid	30							
TREEBARK VT.								
Candleholder on stand	85							
TREE OF LIFE								
(IMPERIAL)								
Bowl, 5½"	27							
Handled Basket	30							
Plate, 7½"	37							
Tumbler	22							
Pitcher	60							
Perfumer, w/lid	40							
Vase Whimsey (From Pitcher)						150CL		
TREE TRUNK								
(NORTHWOOD)								
Vase, 7"-12"	32	45	42	50		985	150IB	
Funeral Vase, 15"-20"	290	320	317	325	900*MO			
Jardinere Whimsey, Rare	290	420						
TREFOIL FINE CUT								
(MILLERSBURG)								
Exterior pattern only								
TRIAD								
Hatpin		35						
TRIANDS								
(ENGLISH)								
Creamer, Sugar or Spooner	48							
Butter	60							
Celery Vase	55							
TRIPLETS								
(DUGAN)								
Bowl, 6"-8"	30	38	42		46			
Hat	26	32	35					
TROPICANA								
(ENGLISH)								
Vase, Rare	1,400							
TROUT AND FLY								
(MILLERSBURG)								
Bowl, 8¾", Various								
Shapes	450	650	595				1,000LV	
Plate, 9", Rare		6,000						

280

Tulip Scroll

Unshod

Victorian

Vining Leaf

	M	A	G	B	PO	AO	Pas	Red
TULIP (MILLERSBURG)								
Compote, 9", Rare	700	700	700					
TULIP AND CANE (IMPERIAL)								
Wine, 2 Sizes, Rare	55						65SM	
Claret Goblet, Rare	65						75SM	
Goblet, 8 oz, Rare	75						80SM	
TULIP SCROLL (MILLERSBURG)								
Vase, 6"-12", Rare	260	400	350					
TUMBLE-UP (FENTON-IMPERIAL)								
Plain, Complete	70						80	
Handled, Complete, Rare	275						295	
TWINS (IMPERIAL)								
Bowl, 9"	36		48				42	
Bowl, 5"	24		30				28	
Fruit Bowl, w/base	58							
TWO FLOWERS (FENTON)								
Plate, 13", Rare	695							1,800
Bowl, 5"-8", Ftd	28	45	50	42				
Bowl, 8", Flat	90	120	148	146			255V	1,000
Rose Bowl, Rare	95		140	147				
Bowl, 8"-10", Ftd	145	175	178	287			375W	1,650
Plate, 9", Ftd	500		550	550				
TWO FRUITS (FENTON)								
Divided Bowl, 5½" Scarce	60	85	100	85			115W	
TWO FRUITS (NORTHWOOD)								
Sugar, Rare				450				
Spooner, Rare				450				
TWO ROW (IMPERIAL)								
Vase, Rare		750						
URN								
Vase, 9"	40							
US DIAMOND BLOCK (U.S. GLASS)								
Compote, Rare	56				75			
Shakers, pr	60							
UMBRELLA PRISMS								
Small Hatpin		26						
Large Hatpin		30						
UNSHOD								
Pitcher	75							
UTILITY								
Lamp, 8", Complete	75							
VALENTINE								
Ring Tray	80							
VALENTINE (NORTHWOOD)								
Bowl, 10", Rare	350							
Bowl, 5", Rare	95	125						
474 VARIANT (SWEDEN)								
Compote, 7"			80					
VENETIAN (CAMBRIDGE)								
Vase, 9¼", Rare (Lamp Base)	1,050		900					
Creamer, Rare	400						500CM	
Sugar, Rare	400						500CM	
Butter, Rare	600						750CM	
VICTORIAN								
Bowl, 10"-12", Rare		350			1,600			
VINEYARD (DUGAN)								
Pitcher	90	350			800			
Tumbler	25	46					265W	
VINEYARD AND FISHNET (IMPERIAL)								
Vase, Rare								600
VINEYARD HARVEST (AUSTRALIAN)								
Tumbler, Rare	100							
VINING LEAF AND VT. (ENGLISH)								
Spittoon, Rare	350							
Vase, Rare	250							
Rose Bowl, Rare	265							

281

Vintage (Fenton)

Violet

Virginia Blackberry

Waffle Block

	M	A	G	B	PO	AO	Pas	Red
VINING TWIGS								
(DUGAN)								
Bowl, 7½"	30	40	40				47	
Hat....................................	40	45					55W	
VINTAGE								
(FENTON)								
Epergne, One Lily, 2 Sizes ...	95	135	145	130				
Fernery, 2 Varieties	47	58	65	65			76	800
Bowl, 10"	40	57	60	60			70	1,200
Bowl, 8"	36	45	48	48		900	60	875
Bowl, 6½"	29	40	42	42			55	
Bowl, 4½"	25	36	38	38			45	
Plate, 7¾"		175						
Whimsey Fernery................		175						
Card Tray	36							
Punch Bowl, w/base............	250	390	410	450				
(Wreath of Roses Exterior)								
Cup...................................	24	30	35	37				
Rose Bowl..........................	48			57				
Compote.............................	38	45	46	42				
Plate, 7".............................	100			200				
Plate, 11", Ruffled...............	180		210	250				
VINTAGE								
(DUGAN)								
Powder Jar, w/lid	55	150		150				
Dresser Tray, 7"x11"...........	78							
VINTAGE								
(U.S. GLASS)								
Wine..................................	40	48						
VINTAGE								
(MILLERSBURG)								
Bowl, 5", Rare....................	500		600	800				
Bowl, 9", Rare....................	600	900	800	3,200				
VINTAGE BANDED								
(DUGAN)								
Mug...................................	30						45SM	
Tumbler, Rare	500							
Pitcher..............................	200	500						
VINTAGE VT.								
(DUGAN)								
Plate..................................	295	385						
Bowl, Ftd 8½"			90				450WS	
VIOLET								
Basket, Either Type	50	60		75				
VIRGINIA BLACKBERRY								
(U.S. GLASS)								
Pitcher, Small, Rare.............				225				
Note: Tiny Berry Miniature								
Tumbler May Match This.								
VOTIVE LIGHT								
(MEXICAN)								
Candle Vase, 4½",								
Rare	350							
WAFFLE BLOCK								
(IMPERIAL)								
Handled Basket, 10"...........	47						60SM	
Bowl, 7"-9"	36							
Parfait Glass, Stemmed	30						42	
Fruit Bowl, w/base..............							200CM	
Vase, 8"-11"	40						55	
Nappy................................							40PM	
Pitcher...............................	140						165CM	
Tumbler	200						270CM	
Rose Bowl, Any Size............	70							
Plate, 10"-12", Any Shape	77						165SM	
Sherbet							35CL	
Punch Bowl	175						220TL	
Cup...................................	18						30TL	
Shakers, pr	75							
Creamer	60							
Sugar	60							
WAFFLE BLOCK AND HOBSTAR								
(IMPERIAL)								
Basket................................	145						160SM	
WAFFLE WEAVE								
Inkwell	85							
WAR DANCE								
(ENGLISH)								
Compote, 5"........................	75							
WASHBOARD								
Creamer, 5½".......................	42							

Water Lily

Wheat

Whirling Star

Wide Panel

	M	A	G	B	PO	AO	Pas	Red
WATER LILY (FENTON)								
Bon Bon	35	45	47	45			50	
Bowl, 5", Ftd.	26	32	34	30			52	750
Bowl, 10", Ftd.	75	90	110	80			120	
WATER LILY AND CATTAILS (FENTON)								
Toothpick Whimsey	70							
Bon Bon	60	72		75				550
Pitcher	320							
Tumbler	95							
Butter	160							
Sugar	100							
Creamer or Spooner	90							
Bowl, 5"	32	50		50				
Bowl, 7"-9"	40							
Spittoon Whimsey, Rare	1,600							
WATER LILY AND CATTAILS (NORTHWOOD)								
Pitcher	410							
Tumbler	95	165		2,700				
WATER LILY AND DRAGONFLY (AUSTRALIAN)								
Float Bowl, 10½", Complete	110	120						
WAVEY SATIN								
Hatpin		25						
WEBBED CLEMATIS								
Vase, 12½"	225							
WEEPING CHERRY (DUGAN)								
Bowl, Dome Base	45	65			195		90	
Bowl, Flat Base	40	75			195		80	
WESTERN DAISY (WESTMORELAND)								
Bowl	45	52			167MO			
Hat	40	48						
WESTERN THISTLE								
Tumbler, Rare	285							
WHEAT (NORTHWOOD)								
Sweetmeat, w/lid, Rare		2,350	2,500					
Bowl, w/lid, Rare		2,000						
WHEELS (IMPERIAL)								
Bowl, 9"	45							
WHIRLING HOBSTAR (U.S. GLASS)								
Punch Bowl, w/base	120							
Cup	20							
Pitcher	250							
WHIRLING LEAVES (MILLERSBURG)								
Bowl, 9"-11", Round or Ruffled	75	100	95				120CM	
Bowl, 10", Tri-Cornered	295	395	350				650V	
WHIRLING STAR (IMPERIAL)								
Bowl, 9"-11"	35							
Compote	55		62					
Punch Bowl, w/base	135							
Cup	20							
WHITE ELEPHANT								
Ornament, Rare							350W	
WHITE OAK								
Tumbler, Rare	300							
WIDE PANEL (U.S. GLASS)								
Salt	40							
WICKERWORK (ENGLISH)								
Bowl, w/base, Complete	235							
WIDE PANEL (NORTHWOOD-FENTON-IMPERIAL)								
Bowl, 9"	40	85					70SM	
Compote	36							
Epergne, 4 Lily, Rare	450	575	700	800		6,000	850W	
Console Set, 3 Pieces	90						110	
Goblet	36							190
Cake Plate, 12"-15"	65	90					100	250
Punch Bowl	90						145	975
Cup	20							100
Lemonade, Handled	28						75W	
Compote, Miniature	36							
Covered Candy	32	42					50	350
Spittoon Whimsey	350							
Vase	28	37	37	40	52		60	650

Wild Rose Lamp

Wild Rose Syrup

Windmill

Wine and Roses

	M	A	G	B	PO	AO	Pas	Re
WIDE PANEL (WESTMORELAND)								
Bowl, 7½"							65TL	
Bowl, 8¼"							62AM	
WIDE PANEL BOUQUET								
Basket, 3½"	65							
WIDE PANEL VT. (NORTHWOOD)								
Pitcher, Tankard	185	240	260					
WIDE RIB (DUGAN)								
Vase	38	42	46	46	60	187	65	
WILD BERRY								
Jar, w/lid	110							
WILD BLACKBERRY (FENTON)								
Bowl, 8½", Scarce	50	65	70					
Bowl, "Maday" Advertising, Rare		900						
WILD FERN (AUSTRALIAN)								
Compote.............................	135	165						
WILDFLOWER (MILLERSBURG)								
Compote, Ruffled, Rare........	1,000	1,200	1,500					
Compote, Jelly, Rare		1,600					2,800V	
WILDFLOWER (NORTHWOOD)								
Blossomtime Exterior								
WILD GRAPE								
Bowl, 8¼"	50							
WILD LOGANBERRY (WESTMORELAND)								
Cider Pitcher, Rare							450IM	
Compote, Covered, Rare							200IM	
Creamer, Rare							150IM	
Sugar, Rare							100IM	
Wine...................................	85							
Goblet					100			
Note: Also Known as Dewberry								
WILD ROSE (NORTHWOOD)								
Bowl, 8", Flat......................	35	45	42					
Bowl, 6", Ftd, Open Edge	30	48	42				395IB	
WILD ROSE (MILLERSBURG)								
Small Lamp, Rare................	900	1,000	1,000					
Medium Lamp, Rare	1,200	1,600	1,450					
Lamp, Marked "Riverside," Very Rare...........................			2,500*					
Medallion Lamp, Rare	1,600	1,900	1,800					
WILD ROSE								
Syrup, Rare	595							
WILD STRAWBERRY (DUGAN)								
Bowl, 9"-10½"	85	135			350			
Plate, 7"-9", Rare		285*						
Bowl, 6", Rare.....................	42	65			120			
WINDFLOWER (DUGAN)								
Bowl, 8½"	36			47			85	
Plate, 9".............................	130			160				
Nappy, Handled...................	50	50			145		70PK	
WINDMILL (IMPERIAL)								
Bowl, 9"...............................	28	37	37				125V	
Bowl, 5"...............................	18	24	24					
Fruit Bowl, 10½".................	32		40					
Milk Pitcher........................	47	165	95					
Pickle Dish	20		45					
Tray, Flat	40		65				75	
Pitcher................................	70	200	150				425	
Tumbler	20	70	40				70	
WINDSOR (IMPERIAL)								
Flower Arranger, Rare..........	85						85IB	
WINE AND ROSES (FENTON)								
Cider Pitcher, Scarce	600							
Wine...................................	50			90		400	100AQ	

Wishbone and Spades

Wreath of Roses (Fenton)

Wreathed Cherry

Zipper Stitch

	M	A	G	B	PO	AO	Pas	Red
WINGED HEAVY SHELL								
Vase, 3½"							95	
WINKEN								
Lamp	95							
WISE OWL								
Bank	46							
WISHBONE								
(IMPERIAL)								
Flower Arranger	85						90	
WISHBONE								
(NORTHWOOD)								
Bowl, 8"-10", Flat	75	100	100	300				
Bowl, 9", Ftd	75	125	100			2,500	1,500PL	
Epergne, Rare	250	350	350					900IB
Plate, 9", Ftd, Rare		375						
Plate, 10", Flat, Rare	295	490	490					
Pitcher, Rare	700	1,100	985					
Tumbler, Scarce	75	135	125				300PL	
WISHBONE AND SPADES								
(DUGAN)								
Bowl, 8½"		120			140			
Bowl, 5"		70			85			
Plate, 6", Rare		350						
Plate, 10½", Rare		1,000			1,100			
WISTERIA (NORTHWOOD)								
Bank Whimsey, Rare							1,000W	
Pitcher, Rare							5,000IB	
Tumbler, Rare							1,000IG	
WOODEN SHOE								
One Shape, Rare	250							
WOODLANDS								
Vase, 5", Rare	75							
WOODPECKER								
(DUGAN)								
Wall Vase	38		75				90V	
WOODPECKER AND IVY								
Vase, Rare			1,500				1,500V	
WREATH OF ROSES								
(DUGAN)								
Rose Bowl,	45	58						
Spittoon Whimsey,								
Rare	55						150AM	
WREATH OF ROSES								
(FENTON)								
Bon Bon	36	42	42	40			58	
Bon Bon, Stemmed	42	48	48	46			64	
Compote	40	46	46	42				
Punch Bowl, w/base	2,000	350	365	375	2,000			
Cup	24	30	35	37	300			
WREATH OF ROSES VT.								
(DUGAN)								
Compote	50	60	60	60				
WREATHED BLEEDING HEARTS								
(DUGAN)								
Vase, 5¼"	90							
WREATHED CHERRY								
(DUGAN)								
Oval Bowl, 10½"	85	135		290	375		225W	
Oval Bowl, 5"	32	38		75	60		56W	
Butter	95	160					195W	
Sugar	70	100					110W	
Creamer or spooner	65	95					100W	
Toothpick, Old Only		160						
Pitcher	200	420					775W	
Tumbler	36	54					160W	
ZIG ZAG								
(FENTON)								
Pitcher, Rare, Decorated	180			400			500IG	
Tumbler, Decorated	36			50			70IG	
ZIG ZAG								
(MILLERSBURG)								
Bowl, 9½", Round or Ruffled	240	325	350					
Bowl, 10", Tri-Cornered	400	550	500					
Card Tray, Rare			750					
ZIPPER LOOP (IMPERIAL)								
Hand Lamp, Rare	425						500SM	
Medium Lamp, Rare	550						550SM	
Large Lamp, Rare	550						500SM	
ZIPPER STITCH (CZECH)								
Cordial Set (Tray, Decanter,								
4 Cordials), Complete	1,200							

Zippered Heart

	M	A	G	B	PO	AO	Pas	Re(
ZIPPER VT. (ENGLISH)								
Sugar, w/lid	47							
ZIPPERED HEART								
Bowl, 9"............................	70	110						
Bowl, 5"............................	37	48						
Queen's Vase, Rare.............	1,500	1,200						
Pitcher (Not Confirmed)	600*	1,500*						
Tumbler (Not Confirmed)	80*	150*						
ZIP ZIP (ENGLISH)								
Flower Frog Holder	54							

Books on Antiques and Collectibles

Most of the following books are available from your local book seller or antique dealer, or on loan from your public library. If you are unable to locate certain [book]s in your area you may order by mail from COLLECTOR BOOKS, P.O. Box 3009, Paducah, KY 42002-3009. Add $2.00 for postage for the first book ordered [and] $.30 for each additional book. Include item number, title and price when ordering. Allow 14 to 21 days for delivery. All books are well illustrated and contain [cur]rent values.

Books on Glass and Pottery

0	American Art Glass, Shuman	$29.95
6	Bedroom & Bathroom Glassware of the Depression Years	$19.95
2	Blue & White Stoneware, McNerney	$9.95
9	Blue Willow, 2nd Ed., Gaston	$14.95
7	Children's Glass Dishes, China & Furniture II, Lechler	$19.95
2	Collecting Royal Haeger, Garmon	$19.95
3	Collector's Ency of American Dinnerware, Cunningham	$24.95
3	Collector's Ency. of Cookie Jars, Roerig	$24.95
7	Collector's Ency. of Depression Glass, 9th Ed., Florence	$19.95
9	Collector's Ency. of Fiesta, 7th Ed., Huxford	$19.95
9	Collector's Ency. of Flow Blue China, Gaston	$19.95
1	Collector's Ency. of Fry Glass, Fry Glass Society	$24.95
6	Collector's Ency. of Gaudy Dutch & Welsh, Schuman	$14.95
3	Collector's Ency. of Geisha Girl Porcelain, Litts	$19.95
5	Collector's Ency. of Hall China, 2nd Ed., Whitmyer	$19.95
8	Collector's Ency. of McCoy Pottery, Huxford	$19.95
9	Collector's Ency. of Nippon Porcelain I, Van Patten	$19.95
9	Collector's Ency. of Nippon Porcelain II, Van Patten	$24.95
5	Collector's Ency. of Nippon Porcelain III, Van Patten	$24.95
7	Collector's Ency. of Noritake, Van Patten	$19.95
8	Collector's Ency. of Occupied Japan I, Florence	$14.95
8	Collector's Ency. of Occupied Japan II, Florence	$14.95
9	Collector's Ency. of Occupied Japan III, Florence	$14.95
9	Collector's Ency. of Occupied Japan IV, Florence	$14.95
5	Collector's Ency. of R.S. Prussia II, Gaston	$24.95
4	Collector's Ency. of Roseville Pottery, Huxford	$19.95
5	Collector's Ency. of Roseville Pottery, 2nd Ed., Huxford	$19.95
3	Coll. Guide to Country Stoneware & Pottery, Raycraft	$9.95
7	Coll. Guide Country Stone. & Pottery, 2nd Ed., Raycraft	$14.95
3	Colors in Cambridge, National Cambridge Society	$19.95
5	Cookie Jars, Westfall	$9.95
3	Covered Animal Dishes, Grist	$14.95
4	Elegant Glassware of the Depression Era, 4th Ed., Florence	$19.95
4	Kitchen Glassware of the Depression Years, 4th Ed., Florence	$19.95
5	Haviland Collectibles & Art Objects, Gaston	$19.95
7	Head Vases Id & Value Guide, Cole	$14.95
2	Majolica Pottery, Katz-Marks	$9.95
9	Majolica Pottery, 2nd Series, Katz-Marks	$9.95
9	Pocket Guide to Depression Glass, 7th Ed., Florence	$9.95
8	Oil Lamps II, Thuro	$19.95
0	Red Wing Collectibles, DePasquale	$9.95
0	Red Wing Stoneware, DePasquale	$9.95
8	So. Potteries Blue Ridge Dinnerware, 3rd Ed., Newbound	$14.95
1	Standard Carnival Glass, 3rd Ed., Edwards	$24.95
2	Standard Carnival Glass Price Guide, 1991, 8th Ed., Edwards	$7.95
4	Wave Crest, Glass of C.F. Monroe, Cohen	$29.95
8	Very Rare Glassware of the Depression Years, Florence	$24.95
0	Very Rare Glassware of the Depression Years, Second Series	$24.95

Books on Dolls & Toys

7	American Rag Dolls, Patino	$14.95
9	Barbie Fashion, Vol. 1, 1959-1967, Eames	$24.95
4	Character Toys & Collectibles 1st Series, Longest	$19.95
0	Character Toys & Collectibles, 2nd Series, Longest	$19.95
1	Collectible Male Action Figures, Manos	$14.95
9	Collector's Ency. of Barbie Dolls, DeWein	$19.95
6	Collector's Ency. of Half Dolls, Marion	$29.95
1	Collector's Guide to Tootsietoys, Richter	$14.95
2	Collector's Guide to Magazine Paper Dolls, Young	$14.95
1	French Dolls in Color, 3rd Series, Smith	$14.95
1	German Dolls, Smith	$9.95
5	Horsman Dolls, Gibbs	$19.95
7	Madame Alexander Collector's Dolls, Smith	$19.95
5	Madame Alexander Price Guide #16, Smith	$7.95
5	Modern Collector's Dolls, Vol. I, Smith, 1991 Values	$17.95
6	Modern Collector's Dolls, Vol. II, Smith, 1991 Values	$17.95
7	Modern Collector's Dolls, Vol. III, Smith, 1991 Values	$17.95
8	Modern Collector's Dolls, Vol. IV, Smith, 1991 Values	$17.95
9	Modern Collector's Dolls Vol. V, Smith, 1991 Values	$17.95
0	Modern Toys, 1930-1980, Baker	$19.95

2218	Patricia Smith Doll Values, Antique to Modern, 7th Ed.	$12.95
1886	Stern's Guide to Disney	$14.95
2139	Stern's Guide to Disney, 2nd Series	$14.95
1513	Teddy Bears & Steiff Animals, Mandel	$9.95
1817	Teddy Bears & Steiff Animals, 2nd, Mandel	$19.95
2084	Teddy Bears, Annalees & Steiff Animals, 3rd, Mandel	$19.95
2028	Toys, Antique & Collectible, Longest	$14.95
1648	World of Alexander-Kins, Smith	$19.95
1808	Wonder of Barbie, Manos	$9.95
1430	World of Barbie Dolls, Manos	$9.95

Other Collectibles

1457	American Oak Furniture, McNerney	$9.95
1846	Antique & Collectible Marbles, Grist, 2nd Ed.	$9.95
1712	Antique & Collectible Thimbles, Mathis	$19.95
1880	Antique Iron, McNerney	$9.95
1748	Antique Purses, Holiner	$19.95
1868	Antique Tools, Our American Heritage, McNerney	$9.95
2015	Archaic Indian Points & Knives, Edler	$14.95
1426	Arrowheads & Projectile Points, Hothem	$7.95
1278	Art Nouveau & Art Deco Jewelry, Baker	$9.95
1714	Black Collectibles, Gibbs	$19.95
1666	Book of Country, Raycraft	$19.95
1960	Book of Country Vol II, Raycraft	$19.95
1811	Book of Moxie, Potter	$29.95
1128	Bottle Pricing Guide, 3rd Ed., Cleveland	$7.95
1751	Christmas Collectibles, Whitmyer	$19.95
1752	Christmas Ornaments, Johnston	$19.95
1713	Collecting Barber Bottles, Holiner	$24.95
2132	Collector's Ency. of American Furniture, Vol. I, Swedberg	$24.95
2018	Collector's Ency. of Graniteware, Greguire	$24.95
2083	Collector's Ency. of Russel Wright Designs, Kerr	$19.95
1634	Coll. Ency. of Salt & Pepper Shakers, Davern	$19.95
2020	Collector's Ency. of Salt & Pepper Shakers II, Davern	$19.95
2134	Collector's Guide to Antique Radios, Bunis	$16.95
1916	Collector's Guide to Art Deco, Gaston	$14.95
1537	Collector's Guide to Country Baskets, Raycraft	$9.95
1437	Collector's Guide to Country Furniture, Raycraft	$9.95
1842	Collector's Guide to Country Furniture II, Raycraft	$14.95
1962	Collector's Guide to Decoys, Huxford	$14.95
1441	Collector's Guide to Post Cards, Wood	$9.95
1629	Doorstops, Id & Values, Betoria	$9.95
1716	Fifty Years of Fashion Jewelry, Baker	$19.95
2213	Flea Market Trader, 7th Ed., Huxford	$9.95
1668	Flint Blades & Proj. Points of the No. Am. Indian, Tully	$24.95
1755	Furniture of the Depression Era, Swedberg	$19.95
2081	Guide to Collecting Cookbooks, Allen	$14.95
1424	Hatpins & Hatpin Holders, Baker	$9.95
1964	Indian Axes & Related Stone Artifacts, Hothem	$14.95
2023	Keen Kutter Collectibles, 2nd Ed., Heuring	$14.95
2216	Kitchen Antiques - 1750-1940, McNerney	$14.95
1181	100 Years of Collectible Jewelry, Baker	$9.95
2137	Modern Guns, Identification & Value Guide, Quertermous	$12.95
1965	Pine Furniture, Our Am. Heritage, McNerney	$14.95
2080	Price Guide to Cookbooks & Recipe Leaflets, Dickinson	$9.95
2164	Primitives, Our American Heritage, McNerney	$9.95
1759	Primitives, Our American Heritage, 2nd Series, McNerney	$14.95
2026	Railroad Collectibles, 4th Ed., Baker	$14.95
1632	Salt & Pepper Shakers, Guarnaccia	$9.95
1888	Salt & Pepper Shakers II, Guarnaccia	$14.95
2220	Salt & Pepper Shakers III, Guarnaccia	$14.95
2141	Schroeder's Antiques Price Guide, 10th Ed.	$12.95
2096	Silverplated Flatware, 4th Ed., Hagan	$14.95
2027	Standard Baseball Card Pr. Gd., Florence	$9.95
1922	Standard Bottle Pr. Gd., Sellari	$14.95
1966	Standard Fine Art Value Guide, Huxford	$29.95
2085	Standard Fine Art Value Guide Vol. 2, Huxford	$29.95
2078	The Old Book Value Guide, 2nd Ed	$19.95
1923	Wanted to Buy	$9.95
1885	Victorian Furniture, McNerney	$9.95

Schroeder's Antiques Price Guide

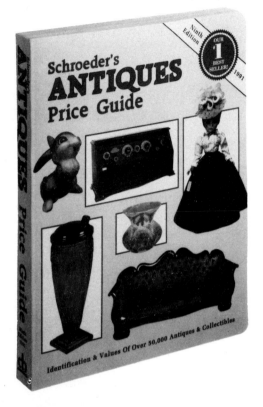

Schroeder's Antiques Price Guide has become THE household name in the antiques & collectibles industry. Our team of editors work year around with more than 200 contributors to bring you our #1 best-selling book on antiques & collectibles.

With more than 50,000 items identified & priced, *Schroeder's* is a must for the collector & dealer alike. If it merits the interest of today's collector, you'll find it in *Schroeder's*. Each subject is represented with histories and background information. In addition, hundreds of sharp original photos are used each year to illustrate not only the rare and unusual, but the everyday "fun-type" collectibles as well -- not postage stamp pictures, but large close-up shots that show important details clearly.

Our editors compile a new book each year. Never do we merely change prices. Accuracy is our primary aim. Prices are gathered over the entire year previous to publication, from ads and personal contacts. Then each category is thoroughly checked to spot inconsistencies, listings that may not be entirely reflective of actual market dealings, and lines too vague to be of merit. Only the best of the lot remains for publication. You'll find *Schroeder's Antiques Price Guide* the one to buy for factual information and quality.

No dealer, collector or investor can afford not to own this book. It is available from your favorite bookseller or antiques dealer at the low price of $12.95. If you are unable to find this price guide in your area, it's available from Collector Books, P.O. Box 3009, Paducah, KY 42002-3009 at $12.95 plus $2.00 for postage and handling.

8½ x 11", 608 Pages   **$12.95**

COLLECTOR BOOKS
A Division of Schroeder Publishing Co., Inc.